T0227525

THE NORTHRIDGE EARTHQUAKE

Who is at risk of disaster?

Are earthquakes natural disasters?

'Natural' disasters have more to do with the social, political, and economic aspects of society than they do with the environmental hazards that trigger them. Disasters occur at the interface of vulnerable people and hazardous environments. People's vulnerability is a result of inequalities in their access to resources and in their exposures to risk, and is compounded through poverty, discrimination, political powerlessness, and other conditions that cause social and economic disadvantage. Disaster reveals particular features of people's vulnerabilities as they struggle to cope with disruptions and loss.

This book concentrates on the social aspects of disaster, focusing on the most expensive disaster to date in US history, the Northridge earthquake of 1994, to examine facets of vulnerability and post-disaster recovery strategies. Surveying the historical and contemporary aspects of life in Southern California the author explains how vulnerability to disaster in California has been shaped by more than a century of immigration, urbanization, environmental transformations, and economic development.

Examining other recent disasters alongside Northridge, this book provides a global view of the social effects of disaster in developed and developing countries. An invaluable insight into the field for students, academics, and professionals in anthropology, geography, sociology, environmental studies, and emergency management, *The Northridge Earthquake* discusses key aspects of sustainable development and state policy and concludes with considerations of ways that vulnerability can be reduced in the future.

Robert Bolin is Professor of Sociology at Arizona State University.

THE NORTHRIDGE EARTHQUAKE

Vulnerability and disaster

Robert Bolin
with
Lois Stanford

Routledge
Taylor & Francis Group

LONDON AND NEW YORK

First published 1998
by Routledge

2 Park Square, Milton Park, Abingdon, Oxon OX14 4RN
711 Third Avenue, New York, NY 10017, USA

Routledge is an imprint of the Taylor & Francis Group, an informa business

First issued in paperback 2016

Copyright © 1998 Robert Bolin and Lois Stanford

The right of Robert Bolin and Lois Stanford to be identified as the Author of
this Work has been asserted by them in accordance with the Copyright,
Designs and Patents Act 1988

Typeset in Garamond by Keystroke, Jacaranda Lodge, Wolverhampton

All rights reserved. No part of this book may be reprinted or reproduced or
utilised in any form or by any electronic, mechanical, or other means, now
known or hereafter invented, including photocopying and recording,
or in any information storage or retrieval system, without
permission in writing from the publishers.

Notice:
Product or corporate names may be trademarks or registered trademarks,
and are used only for identification and explanation without intent to infringe.

British Library Cataloguing in Publication Data
A catalogue record for this book is available from the British Library

Library of Congress Cataloging in Publication Data
Bolin, Robert C.
The Northridge earthquake : vulnerability and disaster / Robert
Bolin and Lois Stanford.
p.   cm.
Includes bibliographical references.
1. Earthquakes—California—Los Angeles Region. 2. Disaster
relief—California—Los Angeles Region. 3. Northridge (Los Angeles,
Calif.) I. Stanford, Lois. II. Title.
HV600 1994.C2B65   1998
363.34'95'0979494—dc21   98–11099

ISBN 978-0-415-17897-6 (hbk)
ISBN 978-1-138-97728-0 (pbk)

CONTENTS

CONTENTS

LIST OF FIGURES

LIST OF TABLES

PREFACE

The discussion and analysis presented here has been influenced by a line of thinking and analysis that has seen little attention among US disaster sociologists. As such it marks something of a departure from my 'usual' research approach. This work concerns the issue of vulnerability to disaster, where vulnerability is understood as a consequence of various kinds of social inequalities. Rather than explaining disaster as caused by some extreme environmental force, a vulnerability approach seeks the explanation of disaster in the ongoing nature of societies and their unequal allocation of risks and resources. While vulnerability analysis, as it has come to be called, has been around for more than fifteen years, my first encounter with it came only a few years ago, and it has significantly altered the ways that I now think about environmental risk and disaster. This book marks an attempt to develop this line of analysis adapted to the particular conditions of one region of the US. To that end this work has several related components: a review of several disaster research traditions, a historical sketch of the dynamics of California's dramatic urban growth and its ecologies of risk, a general framework for an examination of 'First World' vulnerability, and an exploration of the topic in the context of a recent California earthquake. By covering such a range of topics, this book is not intended to be a heavily quantified and exhaustive monograph or metanarrative of the 'entire' Northridge earthquake. It instead investigates conceptual issues while examining evidence of vulnerability through a series of case studies. It has been written in the midst of what will be a long complex process of recovery after the earthquake, thus decisive judgements and conclusions of its effects are elusive.

The book represents the combined research and writing efforts of Lois Stanford and myself. A division of labor of sorts was followed in the writing, with each of us contributing specific sections. Professor Stanford contributed segments of Chapters 4 through 6, primarily those having to do with the qualitative methods, and selected aspects of the case studies. As primary author I wrote the remainder of what appears here and took on the responsibility of smoothing over differences in writing styles, theoretical proclivities, and disciplinary differences of a sociologist and an anthropologist. While I have

ix

attempted to shape the book into a coherent whole by editing, or, and rewriting the manuscript as needed, some idiosyncrasies in writing styles may remain. There was a certain urgency in getting the book written before too many years pass since the disaster, or for that matter, before another earthquake occurred in 'Loss Angeles'. While every effort has been made to include the most recent research on the earthquake, more keeps appearing and at some point it could no longer be included.

Many of the ideas in here are not mine and I owe an intellectual debt to a number of disaster specialists, most of whom I've never met. Who they are will be readily apparent by the frequency of referencing as their influences are revealed. A rough draft of the manuscript as well as sample chapters were reviewed and critiqued by several anonymous reviewers arranged by Routledge. Their comments were both useful and encouraging, as were those of our editor at Routledge, Sarah Lloyd. Of these 'anonymous' reviewers, I offer special thanks to Walt Peacock at Florida International University, for his meticulous critique of what was a very rough draft. Unfortunately both time limits and page limits have prevented me from addressing most of his many useful ideas and insights.

A fond and special thanks is given to Michelle Hegmon at Arizona State University for her good-natured, clear-headed, and systematic critique and editing of innumerable drafts of this work. Whatever value this work may have is likely attributable to her input and it is to her that this work is dedicated. Of course, all oversights, misinterpretations, and other weaknesses, I take full responsibility for.

Bob Bolin
Tempe, Arizona
November 1997

ACKNOWLEDGEMENTS

The research reported here was conducted with the generous support of the National Science Foundation under grant number CMS-9415721, with the two authors as co-principal investigators. However, all findings and recommendations are solely those of the authors and do not necessarily represent the views of NSF. Thanks is offered William A. Anderson, our Project Manager at NSF for his continued support of our modest research activities. Part of this research was conducted with a Research Experience for Undergraduates grant from NSF which allowed us to take two of our top undergraduate students, Allison Crist and Nicole Ordway, into the field to assist in data gathering and processing. Author Bolin would like to thank Dean Wendy Wilkins of the College of Liberal Arts and Sciences, Arizona State University and Robert Snow, Chair of its Department of Sociology for providing him with sabbatical support during 1996, a place to work while initial analysis and writing was completed, and, more recently, a new professorship in that department.

Field research of the type reported on here is dependent on the willingness of people to tell their stories and to answer what must seem like an endless series of questions. We would like to thank all the people at our research sites who generously gave of their time to help us in understanding the Northridge earthquake from local perspectives. Disaster research is intrusive, yet even in the midst of very busy schedules and disrupted lives, our respondents shared their experiences, giving us important insights into their lives in the face of disaster. Among the disaster professionals interviewed, Karma Hackney of the California Office of Emergency Services, deserves special recognition for being generous of her time in providing information and perspectives on managing assistance programs after a large urban disaster. Fred Messick and Andy Petrow were also very helpful in providing documents, maps, and data. Of the many in local government agencies and NGOs who assisted us, we especially thank Kenny Aragon, Bob Benedetto, Adele McPherson, Priscilla Nielsen, Susan Van Abel, Polly Bee, Beth Ferndino, Al Gaitan, Rory Maas, Virgil Nelson, Emigdio Cordova, and Jesse Ornellas. Our graduate assistant Martina Jackson provided important assistance in the conduct of the field research and in segments of the analysis. Figures 1.1, 3.1, and 4.1 were prepared by

Michael Ohnersorgen, Department of Anthropology, Arizona State University. Figures 1.2, 3.2, and 3.5 were prepared by Cherie Moritz at the Arizona State University Geographic Information Systems Laboratory.

LIST OF ABBREVIATIONS

ARC	American Red Cross
CBO	Community-Based Organization
CDBG	Community Development Block Grants
DFO	Disaster Field Office
EDA	Economic Development Administration
EOC	Emergency Operations Center
FEMA	Federal Emergency Management Agency
FRP	Federal Response Plan
GIS	Geographic Information Systems
HCD	California Department of Housing and Community Development
HUD	Department of Housing and Urban Development
HOA	Homeowners' Associations
IA	Individual Assistance programs
IFG	Individual and Family Grants program
IFRCRCS	International Federation of Red Cross and Red Crescent Societies
LAHD	City of Los Angeles Housing Department
LAT	*Los Angeles Times* newspaper
MHRP	Minimum Home Repair Program
NGO	Non-Governmental Organization
OES	California Governor's Office of Emergency Services
SEMS	Standardized Emergency Management System
SHPO	California State Historic Preservation Office
UMCOR	United Methodist Committee on Relief
USGS	United States Geological Survey
VOAD	Voluntary Organizations Active in Disaster

1

A COMMON DISASTER

On 17 January 1994, an unknown fault under Northridge in the suburban San Fernando Valley of Los Angeles, California, ruptured and briefly thrust the ground violently upward. In those few seconds a disaster was set in motion as buildings and highways collapsed, fires broke out, and people fled into the streets. Some fifty-seven deaths have been attributed to the earthquake and thousands more were injured. The magnitude 6.7 temblor was the source of one of the most expensive disasters in US history, with direct losses of at least US $25 billion, and estimated losses now approaching $44 billion (OES 1997). In the three years that followed more than 681,000 applied for assistance from federal and state governments. One year to the day after the Northridge main shock, a magnitude 6.9 earthquake struck the large port city of Kobe, Japan. That earthquake triggered major fires, buildings and highways toppled as soils liquefied, and thousands of wood framed homes collapsed, killing the inhabitants. As damage assessments became available in the aftermath, it became evident that Kobe's Hanshin earthquake far surpassed Northridge in human and economic costs: more than 6,400 died, 300,000 were homeless after the destruction of 136,000 housing units, and economic losses soared passed the US $150 billion mark (Smith 1996; Tierney and Goltz 1996). In one calendar year, two of the core nations of the North were struck by socially and economically disruptive earthquakes, the worst either had experienced in decades. Both disasters are further reminders that tens of millions of people, particularly those living in heavily urbanized areas along the Pacific Rim, may be exposed, at anytime, to large and damaging earthquakes.

While such summary statistics may provide a general sense of the relative severity of a given disaster, they tell us nothing about lived experiences of people caught up in such catastrophes. Nor do they reveal why a physical hazard such as an earthquake becomes a disaster for some people but not others, and for some communities but not others. Nevertheless, there is a media-driven tendency to focus our attention on the extreme physical aspects of disaster and the more gruesome aspects of people's sufferings: What was the magnitude? How many buildings collapsed? How many died? How much

1

did it cost? There is an equal tendency for many academic researchers to consider such events as natural occurrences caused by some exceptional force of nature temporarily disrupting an otherwise 'normal' community. The United Nations, as part of a broad policy initiative, has declared the 1990s as the International Decade for Natural Disaster Reduction (IDNDR), reinforcing a view that disasters originate in nature (Mitchell 1990). Disasters are easily characterized as unfortunate things that happen from time to time to people and their cities. What is missing in this view is any understanding of the ways that political and economic forces create conditions that result in an earthquake having disastrous impacts for *some* people and communities.

Disasters such as Kobe or Northridge reveal things about people, the places they reside in, and their collective histories. They sometimes reveal things about communities that many, in other less turbulent times, would prefer to ignore. The disruptions of a disaster can unmask social inequalities and the injustices that accompany them. Disasters can destabilize entrenched political interests and provide subaltern groups with opportunities and motivations for social change. Too often, however, disasters become the basis for rebuilding social inequalities and perhaps deepening them, thus setting the stage for the next disaster. Disasters, as social processes, are shaped and structured by the sociocultural formations and political and economic practices that exist prior to the onset of a hazard event (Hewitt 1983b). In this sense, disasters are embedded in the flow of everyday life and the social and economic forces that structure people's lives and the places they live. Thus, cultural anthropology and history, as Oliver-Smith (1994) amply demonstrates, can tell us as much about earthquake *disasters* as seismology, and perhaps more. It is from the social, geographical, and cultural history of people and places that we gain glimpses into who is vulnerable, who is at risk, and what may happen to them as a hazard agent triggers a disaster. The global number of disasters has been increasing over the course of the twentieth century (Tobin and Montz 1997), although the aggregate number of natural hazard events (earthquakes, floods, cyclones, wild fires, volcanic eruptions, droughts) generally has not. This suggests that human actions and inactions are involved in the production of 'natural' disasters (Blaikie *et al.* 1994).

Some of the social factors responsible for this increase in disasters seem relatively obvious. World population, now approaching six billion, has tripled in this century (Southwick 1996) and global economic production has quintupled since 1950 (Brown 1996). But sheer numbers tell us little about vulnerability – about who is at risk. Certainly there are four billion more people now variously exposed to environmental hazards than there were in 1900. But if all four billion were securely housed in well-planned communities, appropriately educated, and had access to adequate diets, health care, and well-compensated livelihoods, their vulnerability to calamities might be minimized. Such is not the case, however: many live in desperate poverty, lack secure housing or even a place to live in altogether. Many more are marginalized and subjugated because of their race, gender, ethnicity, religion, and/or age. And, it must be said, most are subject to the vagaries of global capitalism and its relentless search for low wages and cheap resources (e.g. Pred and Watts 1992). The root causes of people's vulnerability to environmental hazards must be sought in these social and political-economic processes that structure their daily lives.

Understanding the various ways human practices, either individually or collectively, are implicated in disasters has both practical and theoretical significance. The practical significance, of course, is that if we understand how routine human practices, in the face of natural hazards, 'cause' disasters, then ameliorative actions can be taken to lessen the severity of future disasters. These actions ultimately must consider the social factors that make people vulnerable if more secure lives are to be achieved. The theoretical and analytical significance is that, as social scientists, we may use disasters to develop a fuller understanding of how human practices interface with physical environments and what the consequences of those interactions are for both humans and their environments (Hewitt 1997). But those theoretical gains are only important if that knowledge is reflexively applied to promote safer lives, and to empower people and organizations to rebuild safer communities.

In what follows, we develop an analysis of disaster that draws upon anthropological, geographical, and sociological approaches to social and environmental issues. By focusing on a specific disaster, the 1994 Northridge earthquake, and considering, in that context, a series of case studies, we examine issues that pertain to societies exposed to hazardous environments. Thus we see 'disaster research' not as an isolated speciality, but as necessarily connected to current developments in the social sciences, including issues in development, environmental sustainability, urban geography, political ecology, and critical social theory. Our vehicle for this approach is a study, largely ethnographic and qualitative, of the experiences of disaster in four culturally and geographically diverse communities in California. These case studies are set in the context of the Los Angeles region, a vast Pacific Rim urban center. The central theme focuses on 'vulnerability' as a signifier of diverse social conditions that make some people and groups more susceptible to harm and loss from hazards than others (e.g. Blaikie et al. 1994). In exploring vulnerability in the First World, our approach eschews the typical large-scale quantitative surveys of disaster sociology, in order to develop an understanding of local history, culture, and social dynamics, and how those factors shape response, recovery, and development strategies. Our concern, then, is with the social, cultural, and spatial features of disasters, their historical contexts, and the numerous strategies and struggles of people and organizations to cope and respond.[1]

In this introductory chapter we raise some preliminary issues about 'natural' disasters and consider the ecology of risk in California, particularly as it is produced by exposure to earthquakes and other environmental hazards.

Throughout this work, we seek connections between the 'common disasters' of California and the experiences of other people and places with environmental catastrophes. These connections can be made through an examination of the political and economic forces that give shape to communities and their environments.

INTERROGATING 'NATURAL' DISASTER

Academically based social research on disasters has been in existence for at least five decades in the United States (Drabek 1986), and it has maintained a relatively strong focus on managing physical hazards. Only in the last fifteen years have some researchers systematically begun to question the common-sensical (and some would say obvious) view that natural disasters are caused by the extremes of nature. Use of the term 'natural' disasters to refer to earthquakes, as well as other geological, hydrological, and meteorological hazard events, implies that human agency has no part in their causation. While human actions generally cannot cause an earthquake in the sense of doing something to provoke fault movement, they are often critically involved in the disaster that can follow a seismic event. In that sense then, 'natural' is an inappropriate adjective to describe such disasters (Hewitt 1997). As Quarantelli (1990: 18) notes, 'there can never be a natural disaster; at most there is a conjuncture of certain physical happenings and certain social happenings'. Similarly Oliver-Smith argues (1996: 303) that in 'graphic ways, disasters signal the failure of a society to adapt successfully to certain features of its natural and socially constructed environment in a sustainable fashion'. While the geophysical dynamics that produce an earthquake are natural phenomena, the disasters that sometimes follow are not.

It has been argued that the strong focus on the physical aspects of a disaster produces a one-sided approach reflected in the ongoing search for engineering and other 'technocratic fixes' for disasters (Hewitt 1995). The physical hazard-centered approach produces a kind of environmental determinism in that the disaster is assumed to be caused solely by natural forces which are largely external to the society in question (Blaikie *et al.* 1994). It directs attention away from the social factors implicated in disaster, including poverty, gender inequality, the lack of entitlements, economic under-development, and ethnic marginalization. Such conditions are endemic problems of everyday life for a large segment of the world's population, and they may become an even greater problem when disadvantaged people are exposed to environmental hazards. Addressing the physical risks of a disaster considers only one component of the risks people face. The development of safer buildings and land use practices should take into account the history, culture, and political economy of people and their settlements if disasters are to be effectively and sustainably dealt with (e.g. Oliver-Smith 1996).

Disasters can affect a broad spectrum of a community, and they may produce losses for wealthy and poor alike. But only very rarely will the wealthy become poor after a disaster, and never will the poor become wealthy. Disasters are seldom democratic, nor are they indiscriminate social levelers as is sometimes claimed. While earthquakes and other natural hazards may cause grave losses for a diversity of people, some of those people are much more likely to avoid or be able to cope with loss and damage. For those who cannot, the accumulated deficits of poverty, discrimination, and other structural constraints can only serve to amplify the effects of the hazard on their lives, deepening the daily struggles they face (Cannon 1994; Maskrey 1993).

As suggested above, some disaster researchers, more than a decade ago, began to articulate a view that natural disasters are better understood as a result of normal and routine aspects of everyday life, specifically, the social, cultural, political, and economic forces that structure and organize the lives of people and the places they inhabit. This initiated an effort to shift attention away from the disaster 'event' toward the social and geographical factors that cause a hazard event to become a disaster. Some analysts began to interpret 'natural' disasters as a depending 'upon the ongoing social order, its everyday relations to the habitat and the larger historical circumstances that shape or frustrate these matters' (Hewitt 1983b: 25). In a seeming inversion of what was 'obvious' about natural disasters, a view has been developed by geographers such as Hewitt that seeks explanations of disaster primarily in the sociocultural and economic features of the societies that are variously affected by natural forces. Their focus has been to develop an understanding of the social structures and material practices that made people more or less vulnerable to environmental hazards. In this approach, the underlying causes of disaster are to be found not in nature, but in the organization of human societies (Varley 1994). The shift in emphasis did not mean abandoning concern with the physical hazards, but rather developing a broader historically informed understanding of the social mechanisms that create unequal risks for people in the face of physical hazards.

This approach received further support with anthropologist Oliver-Smith's (1986) analysis of Peru's devastating 1970 earthquake. Oliver-Smith traced the origins of that disaster through centuries of Peru's colonial and postcolonial history documenting the adoption of unsafe and unsustainable building and land use practices. In his analysis, Peru's seismic disaster was some 500 years in the making, rooted in the complex of economic and political forces that structured development and the human–environment relations (Oliver-Smith 1996). The earthquake and subsequent landslides were simply a trigger for a disaster grounded in poverty, political oppression, and the subversion of previously sustainable indigenous practices (see Chapter 2).

The views advanced by geographers, anthropologists, and other social scientists have come to be referred to as a 'vulnerability approach' to disaster

(Cannon 1994). This approach directs attention to the socio-spatial origins of disaster, and to the effects of social inequalities in producing and shaping disasters. A vulnerability approach has pragmatic implications both for how societies respond to disasters and for how they might reduce the severity of the next one. By searching for the underlying causes of disaster in ongoing social, political, and economic dynamics of a locale, it also helps connect disaster research with much broader theoretical and research issues in sociology, anthropology, and geography.

INITIAL ANALYTICAL CONSIDERATIONS

As social scientists who do research on disasters, we are concerned with the social and spatial dimensions of environmental risk and disaster; specifically how social conditions and practices, broadly defined, produce circumstances that abet the transformation environmental hazards into disasters. While our study focuses on people and places in California, the principles of the analysis can be applied across a range of environmental calamities in a diversity of social contexts. Although mention of California may conjure images of endless suburban prosperity, the state also contains innumerable 'Third Worlds'; ethnic enclaves, migrant labor camps, and impoverished neighborhoods where millions are excluded from an increasingly illusory prosperity, victims of a new regime of flexible accumulation and economic restructuring (Soja 1989; Watts 1992). As Soja (1996b: 445–446) writes, Los Angeles has a marginalized population 'of well more than half a million living precariously in housing conditions little better than those of the worst Third World squatter settlements and shantytowns, [in] . . . what is arguably the most severe urban housing crisis in America'. California also has increasingly impoverished public education and social service sectors, themselves victims of sharply reduced tax revenues in the wake of a conservative 'tax revolt' some two decades ago (Lo 1990). Only California's vast and rapidly growing carceral system – the largest in the US – appears immune from cuts in public spending (Davis 1992). How these political, economic, and social conditions relate to disaster is taken up in Chapter 3. For now, we introduce some conceptual issues and provide a few definitions before turning our attention to the ecology of environmental risk.

Disasters, whether they originate from natural hazards, technological failures, or human produced calamities,[2] are treated here as conjunctural processes. That is, they emerge or develop out of the interactions of environmental forces with the particularities of human settlements and the capacities of people in those settlements to deal with the consequences of those forces. At the same time, there is nothing to be gained by conceptualizing society and nature as static dualisms (*sensu* Giddens 1984), reified and essentialized, separate and independent *things* with clear boundaries and intrinsic properties

(Milton 1996; Plumwood 1993; Soper 1996). Such dualistic thinking appears to underlie the view that disasters are simply the extremes of nature unexpectedly striking unwitting and unprepared human systems (cf. Hewitt 1983b). Once the calamitous event is over, nature – the environment – retreats until the next disastrous intrusion on the otherwise normally functioning society. This obscures both the socially constructed nature of environments (Hannigan 1995) and the social factors implicated in generating disasters. The physical world, discursively represented as 'nature' or the 'environment', is continuously modified and transformed as humans produce the means of their existence across space and time (e.g. Escobar 1996). Political ecology, as a theoretical approach, understands nature as a social production, an object of a technocratic discourse that in turn enables and promotes nature's commoditization and continuous physical transformation through productive systems (Dickens 1996; J. O'Connor 1994).

As we discuss 'natural' disasters, we refer back to the theoretical issues raised by political ecologists (e.g. Peet and Watts 1996) in order to place disasters in the larger context of ongoing environmental transformations. 'The environment' in this sense is more than a mere context for human societies, a static backdrop against which social life is acted out. Rather environments, including their hazards, are seen as an integral part of the social organization of societies, socially constructed and organized, variously constraining or enabling human actions in a continuing socio-spatial dialectic (e.g. Harvey 1989). As Soja (1989: 121) writes, '[t]he space of nature is . . . filled with politics and ideology, with relations of production, with the possibility of being significantly transformed'. We begin from the position that the 'space of disasters' is likewise filled with politics and ideology and class relations.

According to geographers such as Soja (1989), the forms of social landscapes and human settlements are best understood as dialectical expressions of human agency in the structural context of a political economy. If we extend that thinking to disasters, then landscapes, urban or rural, cannot be adequately understood as simple stages upon which disasters inscribe their impacts. David Harvey, in theorizing the shaping of modern capitalist landscapes, reminds us (1985b: 150):

> Capitalism perpetually strives . . . to create a social and physical landscape in its own image and requisite to its own needs at a particular point in time, only just as certainly to undermine, disrupt, even destroy that landscape at a latter point . . . The inner contradictions of capitalism are expressed through the restless formation and re-formation of geographical landscapes.

Urban earthquakes and the disasters that follow, are best grasped as part of a continuous dynamic process, a new factor interjected seismically into the ongoing shaping and reshaping of societies and the spaces they occupy. Marxist and postmodern geographies (Harvey 1990; Soja 1996b) become particularly salient when considering the hazard-prone landscapes of heavily urbanized Southern California. The implications of understanding 'nature' and human landscapes as social productions and material expressions of political-economic forces has gone largely unnoticed in the US disaster response literature (cf. Drabek 1986; Kreps and Drabek 1996; Quarantelli 1989). The preference there appears to treat the setting as a given, a routinely functioning organized 'system' without a history or social dynamics apart from what the disaster itself apparently produced.

Disasters can, on occasion, profoundly disrupt human landscapes, and in the flux that follows entrenched interests sometimes enter into periods of conflict with new political voices seeking alternative resolutions to the crisis (Poniatowska 1995; Schulte 1991). There is adequate evidence, as reviewed in subsequent chapters, that many people in a given locale may well seek a quick return to the status quo ante after a disaster, even if that return makes the next disaster nearly inevitable. Too often a return to what was before the calamity simply reproduces the patterns of vulnerability that generated the disaster to begin with. In some cases, however, there are those who try to challenge existing practices as communities are rebuilt, to make places and people more secure against future risks. These efforts may be opposed and contested, particularly if they threaten patterns of privilege and profit, or disrupt existing social relations and land use patterns (Anderson and Woodrow 1989). In spite of, or perhaps because of, the material losses and human costs they involve, disasters sometimes create political openings in which those with long-standing grievances mobilize and engage in struggles for social change (Laird 1991; Olson and Drury 1997). The outcomes of such efforts are by no means certain and often the protagonists for change must continually contest the historic weight of entrenched interests.

Without adequately considering this complex interplay of human agency and social structure in historical and geographical contexts (e.g. Giddens 1990; Soja 1989), the social factors involved in shaping disasters will be necessarily neglected. If, as Harvey (1985a, 1985b, 1989) demonstrates, the changing socio-spatial patterns of urban areas are concrete expressions of the ever-changing requirements of both production and consumption under capitalism, it becomes increasingly difficult to sustain an examination of urban disasters that does not consider, in some fashion, the history and political economy of the involved areas. Disasters, from a vulnerability perspective, are understood as bound up in the specific histories and socio-cultural practices of the affected people taken in the context of their political and economic systems.

Humans and their settlements face innumerable possible risks from natural agents such as earthquakes, floods, and tropical storms and from technological hazards that include nuclear power plants, toxic wastes, air pollution, and water pollution (Hewitt 1997). Some sociologists have suggested that these risks in late modernity have become so generalized that the term 'risk society' is appropriate to capture the global condition (Beck 1992). However, to suggest that risks are totalized in the 'juggernaut' of modernity (Giddens 1990) obscures the vast inequities in risk exposure, inequities that typically flow along class and ethnic lines (Szasz 1994). In general terms, exposure to these hazards varies widely in time and space and according to people's positionalities by class, race, ethnicity, age, gender, and nationality. It has been established, for example, that toxic waste sites in the US are much more likely to be located adjacent to poor and politically powerless communities of ethnic and racial 'minorities' (Bullard 1990). The same social and cultural factors that produce differential exposures to technological hazards are also implicated in differential exposures to natural hazards, although perhaps lacking the former's intentionality (e.g. Blaikie *et al*. 1994).

The likelihood that a person will be negatively affected by environmental hazards refers to his or her *vulnerability*. Blaikie *et al.*(1994: 9), in the most systematic statement of this approach to date, define vulnerability as:

> the characteristics of a person or group in terms of their capacity to anticipate, cope with, resist, and recover from the impact of a natural hazard. It involves a combination of factors that determine the degree to which someone's life and livelihood is put at risk by a discrete and identifiable event in nature or in society.

It is people's vulnerability to hazardous situations, natural or otherwise, that promotes or inhibits disastrous consequences. Disasters are social processes typically initiated by a specific natural hazard[3] agent 'which leads to increased mortality, illness and/or injury, and destroys or disrupts livelihoods, affecting the people of an area such that they (and/or outsiders) perceive it as . . . requiring external assistance for recovery' (Cannon 1994: 29).

Vulnerability is a social characteristic of individuals and groups and as such it is variable. As we review in the next chapter, an individual's or group's vulnerability varies over time and is different for different hazards. Vulnerabilities are unevenly distributed among individuals, households, communities, nations, and entire regions of the world. The locational and temporal variability of vulnerability makes abstract generalizations difficult and points to the need to examine the specific constituents of vulnerability in concrete settings. To say, for example, that the poor are vulnerable raises questions of: 'To what?' 'In what ways?' 'Why?' Vulnerability concerns the complex of social, economic, and political conditions in which people's everyday lives are embedded that structure the choices and options they have in the face of environmental hazards. The most vulnerable are typically those with the fewest choices, those whose lives are constrained, for example, by discrimination, political powerlessness, physical disability, lack of education and employment, illness, the absence of legal rights, and other historically grounded practices of domination and marginalization (Blaikie *et al.* 1994). Clearly the poor, as a social category, are likely to suffer from many of these constraints and the connections between poverty and vulnerability are typically strong (Hewitt 1997). Such 'clusters of disadvantage' (Chambers 1983: 108) are evidenced in the course of actual disasters, where the elements of vulnerability play themselves out in the lives of people who attempt to cope and to recover their losses.

The term 'vulnerability' itself has been used in more conventional discourses on disasters, but its meaning is usually used to refer to a range of phenomena (Hewitt 1995). Vulnerability is often employed to refer solely to people's *physical exposures* to hazard agents, which shifts attention back to the physical agent, and away from social factors that make their exposure significant or not (e.g. Kreimer and Munasinghe 1992). Vulnerability analysis argues that it is not exposure alone that makes people vulnerable, but in combination with the social and economic factors that constrain their coping abilities. The term's most common use continues to be its application to the quality and safety of structures rather than to the social characteristics of people exposed to environmental hazards. The safety of structures cannot be adequately considered without equal weight being given to the socio-economic characteristics of their inhabitants. It does little good to build 'safe' structures if people cannot afford to live in them or will not live in them because they are culturally inappropriate. An essentialist meaning to vulnerability sometimes appears in cases where it is employed to refer to loosely defined group characteristics. Thus, 'minorities', the elderly, women, and children, are sometimes identified as 'vulnerable groups' (e.g. Aptekar 1994), where vulnerability is used to connote an apparently intrinsic characteristic of certain demographic categories of people. While such groups certainly may prove to have vulnerabilities, essentialist usage elides examination of the particular ways that members of such categories may or may not be vulnerable to hazards. It also fosters interventions that treat people from those categories as necessarily helpless due to their putative intrinsic vulnerability. None of these uses is theoretically consistent with vulnerability analysis as developed in recent work by geographers and anthropologists (e.g. Hewitt 1995; Oliver-Smith 1996). The strength of a vulnerability approach lies in the shift in focus away from the hazard agent to the complex of social, cultural, political, and economic factors that produce different levels of risk for different people across time and space. We recognize that understanding the physical aspects of hazard agents is essential in engineering and other technical interventions that result in safer structures and land uses (Smith 1996). But research on physical hazards is not a substitute for understanding the social processes that deny some access to livelihoods and resources that would allow them to cope with disasters or avoid them all together.

It has long been observed that the scope, magnitude, or intensity of the hazard agent may have little to do with the severity of the disaster that follows (Varley 1994). A magnitude 9.2 earthquake in Alaska in 1964 killed 115 people, destroyed much of the city of Anchorage along with numerous rural villages, producing property losses in excess of $300 million. In response, a massive (for the time) US government rebuilding program was initiated and made significant progress toward reconstruction within one year. In spite of the severity of the physical losses, recovery was deemed complete in three years (Quarantelli and Dynes 1989). In contrast, a magnitude 7.7 earthquake struck the rural highlands of Peru in 1970 triggering the worst disaster recorded in the western hemisphere. In its aftermath, approximately 70,000 were dead, half a million homeless, and the material infrastructure of a large underdeveloped region was destroyed (Oliver-Smith 1994). Recovery, now measured in decades, has been slow, sporadic, and uneven, impeded by vested interests in the power center of Lima and Peru's continuing debt crisis. While thousands of residents suffered loss and deprivation, and struggled to recover after the Great Alaska earthquake, many were able to tap a large, well-financed federal response to assist in their rebuilding efforts. The exceptions to this were rural native villages where poverty and limited access to assistance made recovery more difficult. Overall, the state of Alaska had a robust economy at the time, a factor which contributed to recovery. In contrast, as Oliver-Smith notes (1994: 40), 'Peru was and continues today to be in a disaster condition.' The broad inequalities of power and wealth in Peru derived from its colonial and postcolonial history, combined with deep rural poverty, produced chronic vulnerability among large segments of its rural indigenous population. These vulnerabilities were starkly revealed by the earthquake and the protracted disaster that followed. These brief examples point to a few of the social, political, and economic factors that channel the effects of a physical hazard and structure the course and consequences of a disaster. Earthquakes of the same magnitude can trigger very different disasters depending on where and when they occur, and on the levels of societal protection available.

EARTHQUAKES AND THE ECOLOGY OF RISK

The main subject of this book is a disaster in California, one triggered by the 17 January, 1994 Northridge earthquake. This magnitude 6.7 (M6.7) temblor occurred near the epicenter of the 1971 M6.4 San Fernando (Sylmar) earthquake. Both were centered underneath the suburban sprawl of the San Fernando Valley of Los Angeles County, California (Figures 1.1 and 1.2). The Northridge earthquake is only the latest in a long line of moderate earthquakes in California, a phenomenon as synonymous with the state as

Table 1.1 Losses in recent California earthquakes

Location	Year	Magnitude	Deaths	Injuries	Damage (in $ million)
Northridge*	1994	6.8	57	9,000+	44,000
Landers	1992	7.6	1	402	48.5
Cape Mendocino	1992	7.1	0	356	51.5
Sierra Madre*	1991	5.8	1	30+	36
Loma Prieta	1989	7.1	63	3,757	6,500
Whittier*	1987	5.9	8	200+	430
Oceanside	1986	5.3	1	28	.9
Coalinga	1983	6.4	0	47	42
Livermore	1980	5.5	1	44	17.5
San Fernando*	1971	6.4	58	2,000	1,766
Total			190	15,864	52,892.4

Source: OES (1997).
Note Amounts in 1994 dollars
* Indicates epicenter in Los Angeles County

Disneyland, freeways, and smog. Both Sylmar and Northridge are examples of what we call a 'common disaster', moderate and relatively frequently occurring disasters. Northridge is a common disaster in that its aftermath clearly resembles what has followed other recent California earthquakes, if on a somewhat larger scale. By representing Northridge as 'common', attention is given to the routine aspects of people's everyday lives that made the earthquake a disaster for some but not others. In Table 1.1, a selected list of California's recent damaging earthquakes illustrates the relative scale of Northridge.

What the Northridge earthquake has in common with previous earthquake disasters has more to do with the history and political economy of California, than with particular features of its seismology. California's history has produced ethnic and class divisions that affect the spatial and social distributions of economic and political power in the state (Davis 1992). It is in these social inequalities and their specific local expressions that differences in vulnerability to hazards originate. It is the consistent patterns of social, cultural, and economic differences in vulnerability that give many of the state's earthquake disasters a commonality. As we explore in subsequent chapters, recent earthquakes in California have had uneven effects on people and their communities. In particular, persons who have been variously marginalized by age, class, gender, culture, ethnicity and/or national origin, often appear susceptible to the effects of the state's many disasters (Bolin and Stanford 1991; Phillips 1991). Lack of affordable housing and limited access to recovery resources after the 1989 Loma Prieta earthquake, for example,

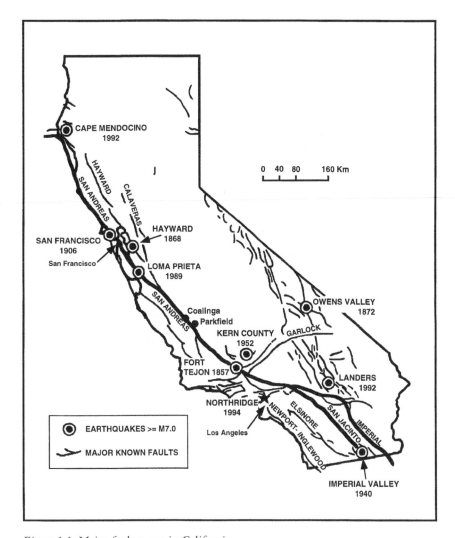

Figure 1.1 Major fault zones in California

compounded the difficulties already faced by many lower income ethnic households on daily basis. Yet vulnerability should not be conflated with helplessness nor does it mean that people do not know what is best for themselves (Kent 1987). In response to new hardships spawned by the Loma Prieta earthquake, poor Latinos in the town of Watsonville organized, mobilized, and ultimately gained new access to resources, housing, and political representation (Laird 1991; Schulte 1991).

The ecology of risk is both spatial and social, involving a mix of geological,

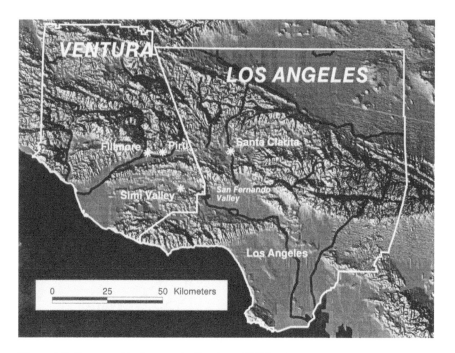

Figure 1.2 Topographic features in Southern California

geographical, and social histories. Much of the human settlement in California is in ecotonal foothill environments that are subject to numerous hazards (Hewitt 1997). Not only is California situated on an active tectonic plate boundary (the San Andreas fault), but it is subject to seasonal extremes in precipitation. It is the only state in the US that has truly seasonal rain with none falling most of the summer and periodic, often severe, winter storms that occasionally produce rain and snowfall better measured in meters than centimeters. One five-day storm in December 1996 produced a meter of rain at one locale and was followed by massive flooding in many areas of Northern and Central California. The dry summer months are also the time of wildfires that sometimes sweep the coastal ranges of the state, occasionally destroying expensive homes perched on hillside 'view' lots as the fires consume the ubiquitous chaparral scrub lands and dry forests. To complete the circle, the firestorms leave denuded hills exposed to the heavy rainfall of winter storms and landslides and mud flows often follow. It should not be surprising that with such persistent hazards every county in California has received at least one federal disaster declaration[4] in this decade alone. Los Angeles County at the time of this writing, has been granted nine federal declarations since 1990, including one for a major urban riot in 1992 (e.g. Soja 1996b). As Joan Didion once remarked, California is 'an amphitheater of natural disaster'.[5] The billions of federal dollars that have flowed to California in the wake of its

14

many disasters reflect the political and economic importance of a state whose gross domestic product, larger than either Italy's or Great Britain's, places it among the top ten nations (Soja 1989).

Southern California's regional history is very much a narrative of economic growth, urban expansion, and the expropriation of land, water, and labor. The growth has been market driven and often unplanned, historically fueled by vast US government subsidies and imported capital (Davis 1992). As a result of convulsive waves of urban expansion throughout the region, from San Diego on the Mexican border north to the high deserts in northern Los Angeles County, a population of more than fifteen million persons now resides in an area routinely troubled by earthquakes, wildfires, and floods (Figure 1.2), to say nothing of some of the worst air pollution in the United States. Of these hazards, earthquakes, while less frequent than floods and fires, have by far the greatest *potential* to produce calamity in the sprawling megalopolis of Los Angeles (Harris 1990).

The estimated risks posed by such hazards are by no means equally distributed across the state or in Southern California specifically. They vary according to spatial factors, geological conditions, and land use patterns that locate some people and settlements nearer hazardous areas than others. The asymmetries of risk are reinforced by social inequalities that pertain in employment, income, wealth, legal protections, and the availability of secure housing. As we have suggested, it is precisely access to these social resources that can influence people's vulnerability to disaster (Blaikie *et al.* 1994; Dreze and Sen 1990). Employment, income, legal status, and ethnicity all bear directly on where people live, the kinds of structures they live in, and whether they have the resources to cope with untoward events in their lives.

In reviewing the ecology of risk in California, it is important to recognize that our understandings of the various risks posed are social products and may or may not mirror the 'real' risks (which are revealed only in an actual disaster). In the case of California's earthquake threat, the estimated probabilities of earthquake occurrence are produced, promoted, and revised constantly by public and private sector geoscientists. As Stallings' insightful (1995) analysis demonstrates, California's earthquake risk has been socially constructed by scientific claims-makers trying to alert an apparently indifferent public to the likelihood of future catastrophic earthquakes. In this sense, the ecology of risk in California is constantly changing, not because the likelihood of earthquakes[6] (or other hazards) is necessarily changing, but because the nature of scientific knowledge is continuously evolving (USGS 1996). Thus, the 'objective facts' of the earthquake risk in California look rather different today than they did twenty years ago, a result of significant changes in theory and research in geology. Most recent estimates of earthquake probability suggest that the risks are increasing as earthquakes are 'overdue' on various fault segments.

California's earthquake problem is the subject of a great deal of ongoing geotechnical and engineering research by an array of academic, private sector, and federal US Geological Survey (USGS) engineers and seismologists. The bulk of this research is federally funded, and findings have wide circulation in academic journals, at congressional hearings, and among state and federal policy making groups (Stallings 1995). This professional community of researchers, engineers, and policy makers, produces innumerable studies on the earthquake hazard in California (and elsewhere) along with policy recommendations for reducing the risk (Office of Earthquakes and Natural Hazards 1993; Palm 1995). Based on an analysis of past earthquakes and on a technically sophisticated research program on current seismic risk, the major thrust of this research has been to forecast earthquakes and design strategies to prepare for them. In particular, it is the threat of 'The Big One', a major catastrophic earthquake either in Los Angeles or San Francisco (or more likely both), that motivates this scientific hunt for understanding and predictability of the physical hazard. The Big One exists both as geoscience prediction and as cultural icon, a looming sword of Damocles invoked by earthquake professionals to promote federal and state regulatory interventions for safer structures and land use planning (e.g. SSC 1995).

With each damaging earthquake, the media immediately, if briefly, raise again the specter of future catastrophe in Los Angeles. For the professional earthquake community, the inevitability of a great California earthquake is no mere cultural icon, the subject of black humor and bad movies. Rather, in their collective view, a future 'great' earthquake that will produce vast losses to property and cost thousands of lives is a virtual certainty, the main questions being only 'where?' and 'when?' (e.g. FEMA 1980; Mileti and Fitzpatrick 1993). The professional earthquake community reasons that by knowing the 'objective' risks of future earthquakes, new earthquake resistant building standards and seismically informed land use planning can be rigorously implemented so that future disaster is avoided (e.g. SSC 1995). Their task as professional scientists and risk managers is to convince a reluctant government apparatus and a naive public of the increasing seriousness of the situation (Palm 1995).

The level of specific scientific knowledge about seismic risks in California is probably unparalleled in the US or the rest of the world. Literally thousands of seismological studies have been produced that map the locations of highest earthquake risk and the tectonic dynamics that produce the threat of earthquakes (Bolt 1993). In mapping these spatial distributions of earthquake risk, estimates are often made as to the likelihood of a given magnitude earthquake occurring by a certain year in different locales in California. This kind of information is used, in turn, to try to change land use planning and building codes to reduce future seismic damage to infrastructure and buildings.

While California's earthquake threat is scientifically well known, it is also a relatively recent threat. Although the west coast of North America has been occupied by humans for millennia, it took specific forms and densities of human habitation in conjunction with particular kinds of structures and economic practices that have produced the threat of earthquake disasters. As Reisner (1993: 501) points out, '[e]arthquakes are quite harmless until you decide to put millions of people and two trillion dollars in real estate atop scissile fault zones'. California has had at least eighty magnitude 6.0 or larger earthquakes (Palm 1995) since European and US colonization began a process of inserting large numbers of people onto the landscape. A number of these earthquakes, including the 1906 San Francisco disaster, have occurred along the well-known San Andreas fault (Figure 1.1), a tectonic plate boundary that is an earthquake threat for both Los Angeles and San Francisco (Bolt 1993). A few of those historic earthquakes have triggered disasters of varying degrees of severity, but none has discouraged the seemingly relentless 'market driven overdevelopment' (Davis 1992: 7) of the state. In fact, water and not earthquakes has been, and continues to be the key environmental factor affecting California's one hundred and fifty years of *laissez-faire* growth (Gottlieb and FitzSimmons 1991), for the availability of water and not the threat of major earthquakes has most strongly affected and directed growth.

The 1906 M8.3 earthquake and subsequent fire that destroyed most of San Francisco is the only *major* earthquake disaster in California's history. While there have been numerous other earthquakes of similar magnitude in the region, only the 1906 earthquake produced such an extensive physical and social disaster. Up to 3,000 died in the disaster and as many as 350,000 persons – virtually the entire city – were left homeless as a result (Harris 1990). The bulk of structural losses in San Francisco resulted from fires that burned out of control due to ruptured water mains. Given the economic forces pushing growth in California it is not surprising that the 1906 earthquake did little to slow the Bay Area's growth. Today more than three million live adjacent to the San Andreas, Hayward, and Calaveras faults which bracket San Francisco Bay (Figure 1.1). The potential for earthquakes emanating from these faults is thought to be an increasingly significant hazard for this area of California (Palm 1995; Palm and Hodgson 1992; USGS 1996).

While Los Angeles has had several earthquakes of some note in this century, none compares with the 1906 San Francisco calamity in loss of life, structural damage, and population displacement. There has, in fact, yet to be a 'Los Angeles earthquake' of the totalizing scope of San Francisco's. The 1933 Long Beach, the 1971 Sylmar, the 1987 Whittier-Narrows, and the 1994 Northridge earthquakes have all occurred within Los Angeles County (Table 1.1). Reflecting the areal extent of Los Angeles (10,515 square kilometers) and its political geography, each earthquake in the urban region has taken its name from the jurisdiction or geologic fault nearest the epicenter.

While much of the state of California has earthquake hazards, the greatest risks of disaster are concentrated in the Los Angeles metropolitan area and the conurbation surrounding San Francisco Bay. According to Comerio *et al.*

(1996) the majority of the state's population lives within 30 kilometers of an active fault zone. Indeed, 40 per cent of the state's population lives in the three counties affected by Northridge (OES 1997). Key in the ecology of risk as currently understood are the myriad of known and named faults that are adjacent to major urban population centers. Los Angeles would appear to face a significant additional risk from a concentration of blind thrust faults (those with no surface manifestations) that apparently underlie much of the metropolitan area. These often unknown faults are a result of the Pacific plate 'bumping into' the North American plate where the plate boundary – the San Andreas fault – has a westward bend in it. The region is thus known as the 'Big Bend' and it produces a compression and faulting of the strata underneath the Los Angeles area as the Pacific plate 'tries' to move north past the North American plate (SSC 1995). The 1994 Northridge earthquake, one effect of this compression, was produced by a previously unknown thrust fault with a hypocenter more than 15 kilometers beneath the San Fernando Valley. The ground movement produced by thrust faults appears particularly destructive to the built environment, subjecting buildings to rapid vertical acceleration (USGS 1996). The faults and the risks they pose are the subject of significant concern among policy makers and earthquake safety advocates (Mileti and Fitzpatrick 1993; Palm 1995). These concerns are based on estimates of earthquakes forecast to occur, with varying degrees of certainty, sometime in the near future. Numerous probable maximum loss (PML) estimates are circulated by earthquake safety advocates inside and outside government agencies as part of their policy making activities (Palm and Hodgson 1992). These worst case scenarios are based on projections of what might happen in a given area should a specific fault rupture releasing a certain magnitude of energy, taking into consideration the types of structures and lifelines present, and population densities (Eguchi et al., 1997). The PML estimates are also an important part of how insurance companies calculate their liability exposure on policies that they have written in the state, a key issue when more than $250 billion in property is covered by earthquake insurance (Roth and Van 1997).

The portraits of future devastation advanced in these loss estimates are hardly encouraging for residents of California and elsewhere along the Pacific Rim. Shah (1995) provides probable loss estimates for great earthquakes in three Pacific Rim urban centers: Los Angeles, San Francisco, and Tokyo. For a repeat today of the 1906 M8.3 San Francisco earthquake (a favorite scenario in these matters), total economic losses are projected at more than US $115 billion. Death tolls, depending on the time of the day the main shock occurs are projected at between 2,000 and 6,000. Other estimates of a repeat of the 1906 earthquake under current conditions predicts $40 billion in property losses (Petak and Atkinson 1985). For Los Angeles, based on a M7.0 event on the Newport-Inglewood fault (Figure 1.1), total economic losses are projected at between $120 and $180 billion. Projected death tolls – between 2,000 and

5,000 – are similar to those for the San Francisco scenario. Both estimates are congruent with actual losses experienced in Kobe, Japan after their 1995 calamity. For Tokyo, a repeat of the M8.2 Great Kanto earthquake of 1923 was modeled (Shah 1995). The numbers produced are illustrative of valuations of property, real or fictitious, in capitalist centers of the Pacific Rim. For the Tokyo metropolitan area, total property losses were estimated in excess of US $2 trillion,[7] with loss of life of at least 40,000 people. If such a computer model bears even a slight resemblance to future earthquake damage in Tokyo (and questions have been raised about them), the financial shocks to Japanese and global capitalism might produce a much broader economic calamity.

In citing such worst case earthquake scenarios for US and Japanese cities, this by no means implies that such wealthy centers of capitalism face the greatest risks of calamity. The cumulative wealth of such countries, the strength of their built environment, and their abilities to absorb disaster losses, make even major earthquakes such as Kobe's 1995 disaster relatively moderate perturbations viewed macrologically. By comparison Guatemala's 1976 earthquake, with losses of around US $1 billion, may appear small in absolute dollar amounts. Yet that M7.9 earthquake triggered a deep social crisis, a disaster that caught up much of the country with 22,000 dead and one million homeless (Anderson and Woodrow 1989). The urban poor and rural indigenous populations suffered a prolonged and deadly shock to their lives and communities, a result of underdevelopment, deep class inequalities, political oppression, and a mismanaged recovery effort (Davis and Hodson 1982). No earthquake in a developed country has engendered such a broad and prolonged disaster.

Significant earthquake risks, as understood by geological scientists, are known to be present in at least thirty-five countries, the majority of which, like Guatemala, lie on or near the so called 'Ring of Fire' of the Pacific Ocean (Smith 1996). An additional zone of high earthquake hazard runs from the Mediterranean Sea and the European Alps across the Middle East and the Himalaya into the Austral-Asia region (Bolt 1993). Because of high population densities, poverty, and unsafe structures, some of the highest losses of life in twentieth-century earthquakes have occurred in areas of Asia and the Middle East. In 1976, the M7.6 Tangshan, China earthquake killed between 240,000 and 800,000 – estimates vary (Hewitt 1997). A similar size main shock (M7.7) in Iran in 1990 left some 40,000 dead and nearly half a million persons homeless (Bolt 1993). The combination of large numbers of vulnerable people, unsafe structures, and frequent earthquake hazards produces disasters in such 'less developed' societies that routinely exceed the worst case scenarios of earthquake disasters in the US and other core capitalist states.

Loss projections, for all their potential weaknesses, represent state of the art estimation procedures for physical risks from earthquakes on known faults. The claims of future losses reproduced above are emblematic of the innumerable 'worst case scenarios' that have been developed over the last two

19

decades in the US. If there is a common theme in these loss exercises (cf. Harris 1990; Palm 1995) it is that the future for earthquake losses in the US and globally is going to be worse (or much worse) than the past. Yet these risk models treat possible future disasters as if they are divorced from the normal and ongoing political and economic processes that create vulnerability to environmental calamity (Hewitt 1983a). For those who rely on such projections, earthquakes are technical problems suitable for a combination of engineering, planning, and specialized managerial solutions, and people, if they are mentioned at all, are seen largely as impediments to carrying out the technocratic solutions, because they fail to see the risks they face (e.g. Mileti and Fitzpatrick 1993).

The focus on the physical hazard agent, particularly in its more violent forms, can tell us something about potential physical risk. It is also clear that, on all accounts, California has a substantial earthquake potential, and that there are places in the state that would appear, given current understandings, to be particularly hazard prone. However, by concentrating on the physical risks, projected extreme events, and worst case scenarios, much is ignored. A vulnerability approach, as reviewed in Chapter 2, directs attention back to people and the common everyday aspects of their lives that make them more or less likely to be caught up in a disaster. While modeling the risks of totalizing 'mega-disasters' may be an interesting exercise, our concern is with common disasters, those relatively frequent events that play themselves out across the fault lines of social vulnerability. Earthquakes the size of Northridge and Kobe occur on average 120 times per year globally (USGS 1996), triggering calamities small and large for people and their communities.

OVERVIEW OF THE BOOK

In this introduction we have raised numerous issues that are addressed in the chapters to come. The value of a vulnerability approach lies in its openness to cultural specificity, social variability, diversity, contingency, and local agency. The research reported here, with its use of ethnographic interviews and other qualitative techniques, lends itself to an examination of local contexts and approaches for dealing with disaster. It is the local struggles and strategies that can provide lessons for dealing with disaster across a range of societal contexts. While presenting an overview of the Northridge disaster is part of this study, a more important part of our project involves providing space for our study participants to speak. Too often disaster research proceeds with the 'view from above', a metanarrative of systemic response and technocratic management, while neglecting the voices of those whose lives are directly caught up in the disaster. In a trenchant critique of conventional disaster research, Hewitt (1995: 330) writes,

To listen to, value, and try to understand the plight and experience of ordinary people in everyday settings, and the victims of disaster, presupposes a concern with who they are and where their experiences take place. To focus on their words is to recognize that these are the only way to recover experiences in other places and times. To pay close attention to what they say, their story and concerns, gives them direct entry into the concepts and discussions of social and disaster research.

The work of Hewitt and others who have developed vulnerability analysis for disaster is presented in Chapter 2, 'Perspectives on Disasters', where we review selected analytical approaches to social research on disasters. We begin with an assessment of several approaches used in US disaster studies. This includes a review of the early 'collective behavior' approach to disasters, the 'systems response' perspective (Drabek 1986), as well as recent developments in 'socio-political ecology' (Peacock et al. 1998). Others have grouped these various approaches into what have been called the 'behavioural paradigm' (Smith 1996) or the 'hazards paradigm' (Hewitt 1995). As Kuhn observes (1970), characteristic of paradigmatic schools is the exclusion, rejection, or marginalization of other, sometimes newer, approaches. Research done in the dominant 'hazards paradigm' tradition has made empirical contributions in US disaster research. At the same time it is this dominant view that some researchers outside the US have been in the vanguard of questioning (e.g. Blaikie et al. 1994: 19).

In the second section of Chapter 2, we review research that provides alternatives to the prevailing hazard-centered approach, beginning with social constructionist issues. We next examine the ethnographic research by anthropologists such as Oliver-Smith and their contributions to situating disasters in larger historical and sociocultural contexts. Anthropologists have provided the research community with needed ethnographic detail about the local struggles and accomplishments of 'victims' in coping with calamity (Schulte 1991). Recent anthropological disaster research complements research by social geographers on disasters and development in Third World[8] settings (Varley 1994). By examining the ways development and underdevelopment produce both social inequalities and vulnerability to disaster, the work by geographers has been foundational in the articulation of vulnerability analysis (Cannon 1994; Hewitt 1983b). After reviewing these somewhat diverse approaches, we illustrate the principles of a vulnerability perspective by reviewing several case studies of earthquakes. While much of the initial work on vulnerability to hazards has focused on countries of the South, we rely on examples from both North and South to illustrate ways that social inequality, political powerlessness, and cultural marginalization produce vulnerability to hazards.

Chapter 3, 'Situating the Northridge Earthquake', examines the historical trajectories of development in California in the context of its environment and resources. Water, particularly the expropriation of distant water sources for Southern California, has been the key resource strategy for promoting growth in the 'Golden State'. A second historical element in shaping California has been the exploitation of low-wage immigrant labor. Immigrants became the backbone of California's giant agricultural system and have provided significant labor pools for much of its large industrial production infrastructure (Davis 1992; Watts 1992). This ethnically diverse working-class and poor population is often vulnerable to the state's natural hazards and equally vulnerable to the vagaries of the state's globalized economy with its continuous restructuring, metamorphoses, and upheavals (Pred and Watts 1992; Soja 1996b). The process of urbanization and economic development in California has experienced only minor interruptions from natural disasters such as the 1906 earthquake in San Francisco. As the state's population ballooned in the first half of the twentieth century, earthquakes and earthquake safety only rarely arose as policy issues. Apart from some minor building code adjustments for seismic safety, little was actively done politically to deal with the state's intermittent earthquakes until well into the second half of the century (May 1985; Palm 1995). Chapter 3 traces the evolution of an earthquake policy to show the difficulties in formulating policies and procedures to deal with uncertain future environmental risks. We consider the market-driven resistance to state attempts to control, in the name of seismic safety, how and where people build in California.

Utilizing a constructionist approach (Hannigan 1995; Stallings 1995), we elaborate on our earlier discussion of how knowledge of earthquake risks are socially produced and promoted in California. We examine risk promotion activities in California and how these efforts have produced a relatively comprehensive set of seismic safety policies designed to reduce earthquake risk in California (Mileti and Fitzpatrick 1993). However, we also suggest that seismic safety, as defined by these experts, is of apparently low salience to many residents of the state. This 'indifference' to the claims of seismic safety experts has produced seismic policies driven almost exclusively by scientific and technical elites, with little public support or opposition (Birkland 1996).

The historical overview of development and earthquake policy provides the background for reviewing the events following the Northridge earthquake. The final section of Chapter 3 describes aggregate features of the 1994 earthquake, looking at general distribution of damages to structures and lifelines, the emergency response, and some political issues in managing the crisis. Federal and state disaster assistance became politicized as more than $12 billion was made available to California. A key issue was exactly where this kind of humanitarian assistance might come from in the federal budget. Anti-welfare conservatives in the US congress attempted to take money for the disaster from other social programs already weakened by deep budget cuts, rather than rely on the ideologically distasteful practice of increasing the federal deficit. It signaled that disaster welfare was no longer exempt from budget scrutiny in the current neoliberal regime. The Northridge disaster became further politicized with the onset of a strident anti-immigrant discourse among California politicians. Much of the discourse centered on state Proposition 187, an initiative on the November 1994 election ballot intended to deny health, education, and other social benefits to undocumented immigrants in the state. Thus political culture began to intrude into the ostensibly humanitarian business of helping people caught up in a disaster.

The 'view from above' of the Northridge disaster developed in Chapter 3 provides the background for introducing the case studies in Chapter 4, 'Situating the Communities and the Research'. The chapter begins with a review of historical factors in the production of vulnerability in the research communities. We situate our analysis in the context of four communities on the margins of Los Angeles: Santa Clarita, Simi Valley, Fillmore, and Piru. Conducting fieldwork is discussed in a review of methodological approaches that we engaged to address local diversity and sociocultural differences among the four communities. We emphasize that communities are not mere locales for research that come in to being with the disaster 'event', objectified by the researcher's gaze. Rather, we examine these communities as places with their own histories, locales that have very different meanings for different people, and where uses of space are socially constructed and politically contested (e.g. Rodman 1992).

The theme of disaster recovery is taken up in Chapter 5, 'Responding to Northridge', where we consider first the overall federal and state responses to the disaster, including reviews of the scope of assistance programs and amounts distributed across the two-county disaster region. Following this overview, we examine first, response issues in the City of Los Angeles, and then return to the four communities and their efforts at mobilizing resources to respond to their losses. A key issue confronted by federal and state responders was dealing with a culturally and socioeconomically diverse 'victim' population within the constraints of a highly bureaucratized assistance system. For a variety of reasons, not all who needed assistance were able to acquire it from federal or state sources, and many more were not able to obtain assistance commensurate with their needs. As Chapter 5 explores, local efforts emerged to help marginalized groups in dealing with their 'unmet needs'. Local civic organizations, non-governmental organizations (NGOs), church groups, and citizen advocacy groups, all mobilized to assist residents deal with their recovery needs. We examine these community-based organizations (CBOs) and NGOs, and the factors associated with their abilities to assist vulnerable groups. Acquiring funds from federal, state, or private sources to support their various programs, and overcoming political opposition to their plans were key elements in their response strategies.

The theme of vulnerability is re-examined in Chapter 5 in reference to local experiences after the disaster. We distinguish between *persistent* vulnerability and *situational* vulnerability. The first refers to a relatively chronic existential condition typical of the very poor, the homeless, and persons marginalized because of culture, ethnicity, gender, or legal status. We use the term situational vulnerability to refer to people whose lives are temporarily made vulnerable to calamity due to illness, short-term unemployment, and other contingencies that make them at risk of additional shocks to their lives such as those produced by disaster. Unlike the persistently vulnerable, the situationally vulnerable may be those in the middle class who have become recent 'victims' of economic restructuring, squeezed out of well-paying jobs yet shouldering a middle-class debt load.

Chapter 6, 'Restructuring after Northridge', expands the analysis of social responses by examining the diversity of ways communities restructured after the Northridge earthquake. We stress the significance of local agency and political activism as instrumental in the acquisition of needed resources. These resources, in turn, were used in a variety of innovative programs to address existing problems, and to prepare for future earthquakes. Opportunities emerged as a result of the temporary availability of federal assistance and these were taken advantage of as local activists and politicians took steps to actively reshape their communities. General issues in social protection for future disasters are reviewed as indicators of the changing socio-political contexts of US disasters.

The final chapter, 'Vulnerability, Sustainability, and Social Change', returns to the general theme of vulnerability and its relationship to post-disaster social processes. A basic contention is that while disasters reveal vulnerability, vulnerability cannot be reduced through *disaster* planning or recovery alone. In this discussion we consider the ways that disasters can be taken as opportunities to rethink existing practices, develop plans to increase equity, and reduce vulnerability. Questions of environmental justice and sustainable development are linked to issues raised in vulnerability analysis to place disasters in the context of contemporary society—environment concerns. As part of this concluding discussion, we consider the ongoing 'fiscal crisis of the state' and its implications for disaster response and recovery funding in the future. The chapter concludes with a discussion of vulnerability reduction based on observations of issues that came out of Northridge and other common disasters.

NOTES

1 Because our interest lies in the social realm, broadly defined, we spend little time discussing the physical aspects of hazards. The physical features and characteristics of natural and technological hazards are the subject of a vast technical literature. For clear summaries the reader is referred to Smith (1996) and Alexander (1993).

2 While technologically produced disasters are clearly human in their origins, we use the term 'human-produced' calamities to refer to war and civil strife. According to the Office of Foreign Disaster Assistance (1990), thus far during the twentieth century nearly half of all disaster deaths have been related to wars. Famines account for another 39 per cent of that total. Proportionately few disaster deaths, in comparison, have been produced through natural hazards.

3 While some hazards, such as earthquakes and volcanoes, are the product of natural processes unmodified by human interventions, other ostensibly natural hazards are less and less 'natural'. The impacts of human activities on global climatic systems, with attendant changes in rainfall patterns, storm frequency, and storm severity suggest that meteorological hazards themselves could be influenced by (un-intended) human factors (e.g. Southwick 1996; Flavin 1997). Flavin (1997) cites evidence that both the frequency and severity of meteorological hazards may be increasing as a result of human-induced climatic change. Similarly human modi-fications of riverine systems, from deforesting and paving watersheds to elaborate levee systems, have taken the 'natural' out of many flood hazards (e.g. Smith 1996).

4 Federal disaster declarations are made at the discretion of the President of the United States upon receiving requests from the governor of a given state. If con-ditions warrant it, a declaration is made and the Federal Emergency Management Agency begins assistance to the stricken state. Declarations make available federal monies, in the form of grants and loans, to assist individuals, businesses, and local and state agencies in response and recovery from the emergency.

5 The statement is quoted in M. Reisner's *Cadillac Desert*, on p. 500.

6 Recently claims have been advanced that Southern California has over the last few decades experienced fewer earthquakes than normal and actual frequency may well increase in the future (USGS 1996). The implications of this were the subject of an hour long Public Broadcasting System television show (1995) ominously entitled *Killer Quake*. The program, featuring top geological scientists as expert commentators, coupled with ominous music, promoted a view that the risk of more earthquakes (and *ipso facto*, more disasters) in Southern California was a virtual certainty. Further, the claim was presented that the greatest risks (now) come from a series of recently discovered blind thrust faults scattered throughout the metropolitan area of Los Angeles, and not from the San Andreas fault to the north. Prominently featured among the new threats was the Elysian Park fault which runs under the towering skyline of downtown LA.

7 On a practical level, to project losses from an earthquake in Tokyo that equal nearly one-half of Japan's gross domestic product raises significant questions about the 'realism' of the loss estimation model (e.g. Wiggins 1996). Since such worst case scenarios are used in claims-making activities regarding earthquake hazard reduction, they bear close scrutiny in terms of who is developing the scenario and whether they have particular interests in either high or low estimates of future loss. It takes little to imagine that land developers and real estate interests in California may wish to de-emphasize future catastrophes lest people stop moving to the state and buying homes. Engineering firms offering risk reduction services might, on the other hand, have some interest in high-loss worst case scenarios that could create higher demand for their hazard mitigation services. Academic scientists have an ongoing interest in securing funding for more research and thus

may be expected to make the claim that further research is needed to improve the accuracy of estimates.

8   In this book we rely on the terms 'Third World' and 'First World', interchangeably with the terms 'South' and 'North' (as in hemisphere) to distinguish between those countries with lower levels of industrial development, comparatively low gross domestic products and per capita incomes, and high levels of foreign debt and those that have higher GDPs and per capita wealth. The Second World of the so-called socialist bloc has shrunk dramatically with the decomposition of the Soviet Union, and thus we will have little use for the term. Our usage approximates Wallerstein's (1992) use of 'core' and 'periphery' to designate the hegemonic capitalist states of the North and the exploited dependent postcolonial states of the South. Indeed, it is precisely the exploitation of the South by the North that gives the distinction its meaning. While some use 'more developed' and 'less developed', or more inaccurately, 'developed' and 'developing' to refer to these spheres, we generally do not. We try to avoid any implication that states of the South are somehow following the same path to 'modernization' or development that the core Northern states did centuries ago. At the same time the risk of using any sweeping designation like First and Third Worlds, is that it creates the impression of unified spheres and obscures diversity and difference. We recognize the diversity of cultures, political systems, economic systems, and social practices within and between these broad categories. Indeed, the economic and social variations within a given Third World country may be greater than between First and Third Worlds (Hewitt 1997). As Northern authors, we try to avoid constructing the South as the 'Other' in the matter of environmental hazards. Processes of globalization have only added complexity to this diversity while creating even greater economic polarizations (e.g. Pred and Watts 1992).

2

PERSPECTIVES ON DISASTERS

Disasters are fundamentally social phenomena; they involve the intersection of the physical processes of a hazard agent with the local characteristics of everyday life in a place and larger social and economic forces that structure that realm. We recognize that there can be no 'neutral' observations of a disaster's social effects that are somehow free of the interventions of social theory or the positionality of the researcher. In order to grasp better how disasters are constructed in social science discourse, we map out the diverse intellectual terrain of hazards and disaster studies and characterize some of the main conceptual features of selected social science approaches. We do not attempt an exhaustive summary of research findings or intellectual history (e.g. Drabek 1986; Quarantelli 1987, 1992; Whyte 1986). Rather, our goal is to identify major analytical approaches in historical context to provide background for a discussion of vulnerability analysis.

In keeping with the increasing complexity of hazards and disasters as environmental, technological, and social phenomena, the field of disaster studies is today both theoretically and empirically diverse. In characterizing social research on hazards and disasters, Smith (1996) suggests that the literature can be divided into two general approaches, behavioral and structural 'paradigms' (cf. Hewitt 1997; Varley 1994). The former conceives of disasters as events caused by physical hazard agents and views human behaviors primarily as responses to the impacts. It emphasizes the application of science and technology, usually directed by government agencies and scientific experts, to restore order and control hazards. Elements of this 'dominant view', as Hewitt (1997) refers to it, appear with some frequency in US disaster research, reflected in its ongoing concern with defining unique features of disasters and how they *differ from* other types of social phenomena (e.g. Kreps and Drabek 1996; Quarantelli 1995). In contrast, the structural paradigm stresses various political and economic factors which unequally place people at risk to hazardous environments. In this view, disasters are not discrete events but are *part of* the larger patterns and practices of societies viewed geographically and historically. This structural approach encompasses much of the recent vulnerability work by anthropologists and social geographers (Blaikie *et al.* 1994; Cannon 1994), and traces its roots in the publication of Kenneth Hewitt's edited volume, *Interpretations of Calamity from the Perspective of Human Ecology* in 1983. Structural approaches tie the study of disasters to more general work on society/environment issues and draw from conceptually richer theoretical traditions than those that view disasters as unusual events requiring their own specialized theory.

As it currently stands there is a limited interchange between scholars working within a vulnerability approach (mostly in Third World applications) and those following the dominant (US) approaches. Very few US disaster researchers are included in the encyclopedic bibliography of *At Risk* (Blaikie *et al.* 1994), and the authors are mostly critical of the few who are discussed. Conversely, most US reviews of disaster research,[1] cite little of the vulnerability literature apart from an occasional mention of Hewitt's 1983 work (e.g. Drabek 1986; Quarantelli 1992). Our concern in this chapter is to illustrate some points of convergence between the two approaches, and to offer a preliminary sketch of how a vulnerability framework may be applied in First World settings.

MAJOR APPROACHES IN US HAZARDS AND DISASTER STUDIES

An important characteristic of US disaster studies today is the interpretation of disasters as events delimited temporally and spatially that disrupt the 'normal order' of things (Kreps 1995). There are numerous theoretical and taxonomic models that define disasters and delineate the appropriate subjects and boundaries of disaster research (Quarantelli 1995); many incorporate physical hazard characteristics (scope, speed of onset, duration of impact) as well as disaster sequences (warning phase, impact phase, emergency response) (Barton 1970; Haas *et al.* 1977; Kreps 1995; Tobin and Montz 1997). The general approach to disasters – focusing on the social consequences of, and responses to acute impacts – is not unrelated to the historical development of the field in the context of US military and civil defense concerns in the immediate post-World War II era (Meyers 1991).

Although the human calamities of war are generally not considered in US disaster research today (cf. Hewitt 1997), the roots of US disaster studies can be traced to the Strategic Bombing Surveys of World War II (Kreps 1995).[2] These studies, which included social analysis of the US nuclear attack on Japan, examined civilian 'morale' during sustained military attacks. This approach carried over into the Cold War era of the 1950s, when, using military funding, civilian disasters were studied as small-scale analogs to nuclear war, events whose understanding would lead to better civil defense preparations and better control of civilian populations. This civil defense/ military preparedness mentality of 'command and control' during civilian crises still characterizes disaster management organizations and agencies in numerous national settings (Neal and Phillips 1995; Quarantelli 1997). Whether the US military was able to use findings from these studies is unclear with Quarantelli (1992) maintaining they were of little use to their army sponsors. On the other hand, Fritz, one of the key figures in the research, wrote that the studies sought information that could be used for the 'maximal amount of disruption to the enemy in the event of a war' (1961: 653), suggesting clear partisanship. One consequence of the 'military connection' in US disaster studies is that the research began to focus on large-scale, sudden-onset 'events' delimited in time and space. Quarantelli (1987) speculates that US disaster studies might have developed differently if disasters had not originally been viewed as analogs of military attacks on urban populations.

Many of the federally funded disaster studies in the 1950s were carried out by sociologists and social psychologists, ultimately leading to 'disaster research' becoming a recognizable academic specialization (Dynes *et al.* 1994). A 'sociological perspective' on disaster was developed through a series of studies funded by the Army Chemical Center and conducted by the National Opinion Research Center (NORC), a research institute at the University of Chicago (Drabek 1986; Quarantelli 1987). Many of the sociologists who worked on the NORC studies (concluded in 1954), became members of the US National Academy of Sciences Committee on Disaster Studies (1951–1957) and their Disaster Research Group (1957–1962) (Mitchell 1990). This work attracted sociologists who worked primarily within a subfield of symbolic interaction theory known as collective behavior (Blumer 1971; Quarantelli 1994). Federal funding of disaster research continued through the Office of Civil Defense, and led to the establishment of the Disaster Research Center (DRC) at Ohio State University in 1963 by sociologists Quarantelli and Dynes (Dynes and Drabek 1994). The DRC (now at the University of Delaware) was the first of several disaster research centers in the US, establishing that country as an early leader in disaster and hazards studies (Quarantelli 1991).

While sociologists focused on the social responses to disasters, US geographers concentrated their research on environmental hazards (Mitchell 1990). Gilbert White's work was central in establishing natural hazards studies in geography some five decades ago (Hewitt 1997). When major engineering interventions failed to prevent floods along major US rivers during the 1930s, White (1945) turned his attention to the human factors that create or modify natural hazards such as floods. Social geographers, drawing on human ecology, began to consider human factors to understand patterns of exposure to hazardous environments (Whyte 1986). Hazards research in the US has, from its origins, had a strong focus on informing public policy and hazard management strategies (Mitchell 1990). As Hewitt (1997: 13) comments, '[Gilbert White] placed humanitarian concern at the heart of our professional responsibilities, and he wanted technical and managerial considerations to follow from, and be judged in terms of, that.' White was instrumental in establishing the Natural Hazards Research and Applications Information Center at the University of Colorado, Boulder, and he co-authored, with a sociologist, the first major assessment of hazards and disaster research, a work that brought together much of the sociological and geographical research available to that time (White and Haas 1975). The interdisciplinary work started through the Colorado center has contributed to the interdisciplinary blending of hazard and disaster research in the US.

As the natural hazards field developed over the last five decades it has diversified considerably, although there is a continuing interest in human choices regarding hazardous environments. Prominent in current hazards research is reliance on a 'bounded rationality' model, which posits that humans place themselves in harm's way typically because of inadequate understanding or information regarding the risks (Burton *et al.* 1993). Issues of hazard information, bounded rationality, and risk perception are prominent features of studies of hazard awareness and warning response studies (e.g. Lindell 1994; Slovic 1986; Stoffle *et al.* 1991). Some have criticized this emphasis on 'rational choice' for neglecting the historical, economic, and cultural constraints that structure people's perceptions and limit choices that people can make in their daily lives (e.g. Hewitt 1997; Kirby 1990).

Perspectives in US disaster sociology

Most published work in US disaster sociology can be grouped into three broad analytical areas: collective behavior (Dynes and Tierney 1994), social problems (Kreps and Drabek 1996), and systems response[3] (Drabek 1986). A common thread in all three approaches is the considerable empirical attention given to human behavioral responses before and after hazard impacts occur. The strong applied focus in much disaster research inhibits the formation of elaborate theoretical statements. Common in US studies are data-driven statistical models of limited phenomena or taxonomies and typologies of various aspects of disasters (Bolin and Bolton 1986; Kreps and Drabek 1996). The specificity of disaster theory often isolates it from more general developments in social theory and reflects a tendency to conceptualize disasters as something apart from normal social and cultural processes (Varley 1994).

Quarantelli (1994) asserts that collective behavior theory (Blumer 1971), is the key theoretical tradition in US disaster studies. While a number of subject areas (including crowd behaviors and mass movements) have been included under the collective behavior rubric, its key feature is a focus on non-institutionalized, spontaneous, and unplanned forms of mass social behavior that occur when the 'social order' is disrupted (Gusfield 1994). In a collective behavior framework, disasters are viewed as events that strain or disrupt established behavioral norms (Barton 1970; Form and Nosow 1958). 'Unique' post-disaster behaviors are understood to be guided by 'emergent norms' that provide the ideational framework for a new type of conformity, albeit potentially a short-lived one (cf. Stallings 1994; Turner and Killian 1987).[4] As virtually all disaster studies are *post hoc*, assertions of a 'new normative order' are usually made in the absence of any substantive evidence of the 'old normative order', and the literature is unclear if emergent norms are truly new or adaptations of already existing norms and values to new situations (e.g. Drabek 1986: 119).

In light of theoretical developments in structure and agency theory (Archer 1996; Bourdieu and Wacquant 1992; Giddens 1984), the ongoing use of 'norms' as an explanation of social action in collective behavior studies seems parochial. It discounts the reflexivity and knowledgeability of actors, does not readily account for movements of political resistance and change after disasters, and reduces explanations of complex social action to normative conformity. In the US literature, only Ino Rossi's (1993) ambitious analysis of the 1980 southern Italian earthquake engages contemporary issues in structure and agency theory (see also Palm 1990). As Stallings (1994: 27) has commented, 'there remains too much unexplored territory in the no-man's land between collective behavior and other forms of interaction . . . we [sociologists] continue to occupy trenches and fortifications laid down . . . nearly a century ago'.

Interpretations of disasters as disruptions to the normal order also prevail among those who use 'functionalist' system response and social problem approaches. The telling critiques of functionalism that led to its abandonment in most contemporary theory are well known and need no review here (e.g. Giddens 1984; Gouldner 1970; Habermas 1987; Lemert 1995). Such critiques notwithstanding, functionalist theory, with ideological underpinnings of system order, equilibrium, and its teleology of system needs still has its proponents in disaster sociology. Kreps and Drabek (1996: 136) argue *contra* Quarantelli that much disaster sociology is functionalist. The definition of disaster offered by Kreps (1995: 256)[5] illustrates these functionalist assumptions: 'A disaster is an *event* concentrated in time and space, in which a society or one of its subdivisions undergoes physical harm and social disruption, such that all or some essential functions of the society or subdivision are impaired' (italics in original). This definition involves the teleological assertion that societies have (unspecified) 'essential functions', that, once impaired by an external force, must be restored if the system is to 'survive'. It leaves little room for a discussion of the two most deadly disasters of the twentieth century: famine and war. There is still a ring of truth to Hewitt's (1983b: 28) charge, made some fifteen years ago, that the 'dominant view [of disasters] is unashamedly indifferent to history and to human and environmental diversity'.

The functionalist emphasis on social order and system integration is prevalent in the 'systems response' approach exemplified in a major codification of the literature by Drabek (1986). The approach discusses disasters in terms of how functionally interconnected levels of social systemic complexity – individuals, groups, organizations, communities and societies – respond to external physical forces. For example, when disasters strike 'the family', conceptualized as a 'group-level' system, it is said to 'expand its protective role, reassuming traditional functions . . . ' Dynes and Drabek 1994: 13). Difficulties in defining what constitutes a family and under what circumstances families act cooperatively (Shiva 1989; Wong 1984) are elided. 'The family', in all its diversity, is nostalgically presented as a unit with traditional (patriarchal?) functions, and attention is diverted from its historically grounded social inequalities by gender, age, race, class, and ethnicity (Bolin *et al.*, in press). This approach appears anachronistic in light of recent analysis of gender and family (cf. Wolf 1991), and it ignores feminist approaches to families as social constructs that unfold dynamically in the microprocesses of gender relations (Greenhalg 1995). The notion that families have 'traditional functions' itself is a residue of a 'sex roles' view of the family common in 1950s sociology, now widely abandoned in the gender literature (Connell 1993; Farganis 1995; Jackson 1993; Jackson 1994).

Functionalist theory also appears in statements of disasters as 'non-routine social problems' (Drabek 1989; Kreps and Drabek 1996). This approach relies on a commonsensical argument that objective conditions in the world that people are disturbed by (e.g. crime, divorce, disaster) constitute social problems. Kreps and Drabek argue that because disasters produce objective disruptions in the world, they are thus social problems. However, such arguments beg the question of who decides when something is a 'problem' and at what threshold level (Spector and Kitsuse 1987). In contrast to functionalist arguments, social constructionists (e.g. Hannigan 1995) contend that social problems are socially produced through organized claims-making activities that call public attention to a putative condition that warrants public action. Not all such conditions are successfully constructed into 'full blown' social problems and objective conditions may have little to do with the levels of public concern about a problem (Stallings 1991, 1995). Indeed natural disasters, far from being a consensus social problem, appear to gain, at best, only episodic interest outside of a specialized community of professional risk managers and researchers.

It should be recognized that elaborations of social theory are of secondary concern in the often applied field of disaster studies. Given the importance of federal funding in setting research agendas, US research strives more to inform policy and technical disaster management practice than to advance social theory. As such it maintains a narrow view of disaster as externally caused, a disruption of the normal order, and an event requiring the interventions of technical experts from the unimpaired outside system (Hewitt 1997: 351). While theoretically conservative, the US research community produces a large number of empirical studies that are used to inform government policy. Extensive research and conceptual elaboration have been accomplished in the areas of hazard warning (e.g. Mileti and Fitzpatrick 1993), emergency response (e.g. Scanlon 1994b), evacuation behavior (e.g. Drabek 1994), and emergency management (e.g. Perry 1985). There has been some limited attention to issues in political economy as well (Tierney 1989), although it is a peripheral concern. A growing number of studies look at social responses to technological hazards and disasters (e.g. Kroll-Smith and Couch 1991a; Picou *et al.* 1997), although fewer have been published on long-term recovery issues, social change, or socio-political conflict (e.g. Blocker *et al.* 1991; Bates and Peacock 1993; Olson and Drury 1997). In the last decade there has been more research on social diversity and social inequality in disasters by class, race, age, and ethnicity (Bolin and Bolton 1986; Perry 1987; Phillips 1993). Gender, long neglected in the published research, has recently emerged as a topic of interest (Fothergill 1996; Morrow and Enarson 1996). Of these substantive areas, the examination of structured social inequalities is the most direct connection between vulnerability analysis and the major US approaches.

Addressing shortcomings of more traditional approaches while incorporating issues common to vulnerability analysis, Peacock and his colleagues have offered a comprehensive statement of an ecologically informed approach to disaster (cf. Kroll-Smith and Crouch 1991b). The approach, given the neologism 'socio-political ecology' by Peacock and Ragsdale (1998), involves conceptualizing human settlements as an ecological network of interrelated 'social systems' that, together with their biophysical setting, form an ecological field (Bates 1997). This field encompasses various processes of competition and conflict among the component self-referential systems (organizations, households, political entities), which are linked in numerous contingent relationships (Bates and Pelanda 1994). Although similarly named, socio-political ecology is theoretically distinct from Marxist and post-structuralist approaches labeled political ecology and feminist political ecology (e.g. M. O'Connor 1994; Peet and Watts 1996; Pred and Watts 1992; Rocheleau *et al.* 1996).

Peacock and his colleagues (1998) have given their approach a detailed conceptual explication through a series of studies done after Hurricane Andrew struck Miami, Florida in 1992. They review both the history and urban ecology of Miami and reveal how existing patterns of residential segregation and economic marginalization for different racial/ethnic groups structured the course and consequences of the disaster and recovery. Their approach also examines the larger political economic forces shaping recovery and local attempts at preserving communities and assisting those with unmet needs (Peacock *et al.* 1998). The research addresses a number of themes that are also significant in vulnerability analysis: issues of gender, race/ethnicity, immigration, and social class (Enarson and Morrow 1998). On our reading, socio-political ecology represents a move to a more context-sensitive approach that considers the importance of political-economic factors in shaping the particular vulnerabilities of different groups, and avoids treating disasters simply as extreme events or non-routine social problems.

ALTERNATIVE ANALYTICAL APPROACHES

The last twenty years have seen broad changes in theoretical approaches in the social sciences (e.g. Best and Kellner 1991). Some have argued that theoretical innovation is necessary to account for recent global social, cultural, and economic transformations coupled with a growing ecological crisis (Jameson 1991). The move to 'post-Fordist' flexible regimes of production, the rapid growth of low wage service sector employment, and the decline of the traditional industrial working class and formerly stable middle-class households are creating major social and economic shifts in core industrial states of the North (Smith and Wallerstein 1992). The combination of rapid cultural change, the shifting ethnic composition of populations in the North, the collapse of the Soviet bloc, the changing statuses of women, the growth of transnational corporate power, continued 'hot' wars and population displacements in the South, have all conspired to produce what some would consider epochal shifts toward what have been labeled postmodern conditions (Best 1995; Best and Kellner 1997; Harvey 1990).

The cultural, economic, and environmental metamorphoses that have animated so much recent social research and theory have gained only passing mention in US disaster and hazards research (e.g. Morrow and Enarson 1996; Peacock *et al.* 1998). It appears that recent intellectual developments in the social theory regarding social power, gender, cultural change, and the environmental crisis (e.g. Best and Kellner 1997; Dickens 1996; Foucault 1980; Haraway 1989; Lutz 1995; Rosaldo 1993) scarcely penetrate the self-referential world of disaster studies. Too often, it seems disaster specialists are more concerned with how to 'classify and partition off the issues involved in the subject' (Hewitt 1983b: 13), than to link disasters to larger issues in the social sciences, specifically the growing literature on society–environment transformations (cf. Peet and Watts 1996). While not suggesting that disaster sociology embrace poststructuralist, constructionist, or critical social theories *tout court*, a critical engagement could lead to more productive connections with current interdisciplinary work on environment and development issues (e.g. Pred and Watts 1992).

Vulnerability analysis offers connections by considering disaster within the broad context of history and social relations, rather than as a unique subject requiring a limited and specialized theory. In the first sustained statement of vulnerability analysis, Hewitt specifically criticized the disciplinary isolation of traditional approaches to disasters and hazards: 'The geography of disaster is an archipelago of isolated misfortunes. Each is seen as a localized *dis*organization of space, projected upon the extensive map of human geography in a more or less random way . . . ' (1983b: 12). To counter this 'disaster archipelago' view, Hewitt and others conceptually situated disasters as one effect of many in larger political economic and ecological processes associated with capitalist development and underdevelopment (e.g. Maskrey 1994; Susman *et al.* 1983; Watts 1983, 1992). These social geographers' work was key in shifting the focus away from physical processes in nature and faulty locational decisions by humans. Instead, attention was directed toward social structural factors and their specific expressions in the inequitable conditions that affect the daily lives of people. This conceptual approach also turned the focus away from social *responses to* disaster and towards an examination of the complex political economic processes that produce vulnerable populations, degrade environments, and contribute to the *causes of* disaster. Vulnerability, as an ongoing process, is produced within a web of changing contingencies that contribute toward the level of risks faced by a given household or community (Wisner 1993). As Hewitt notes (1997), it is not a static condition or a 'social problem' needing professional attention to fix.

The geographers who developed vulnerability analysis drew from theoretical traditions usually not considered by US disaster researchers – critical political economic theory (Watts 1983). While having different theoretical bases and foci, vulnerability analysis does relate those aspects of the US research tradition that have given increasing attention to social inequalities, housing problems, poverty, race/ethnic issues, gender, community-based movements, and development issues (e.g. Fothergill 1996; Peacock *et al.* 1998; Phillips 1993). Another significant addition to the research base is Stallings' (1995) social constructionist analysis of earthquake risk.

The social construction of environmental risk

While social constructionism is a broad label that can be applied across a range of theoretical concerns, it generally encompasses those theories that emphasize language, discourse, and the production of meaning as expressions of human agency (Dickens and Fontana 1994; Giddens 1984; Milton 1996). Social constructionists contend that the risks posed by various social or environmental conditions are socially produced and discursively constituted (Hannigan 1995). The origins of social constructionism can be traced to symbolic interaction theory in the 1920s (Blumer 1971) and more recently to phenomenological approaches such as Berger and Luckmann's 1967 classic *The Social Construction of Reality*. As applied to environmental risks, one of the first constructionist statements was made by Douglas and Wildavsky in a 1982 essay on the cultural selection of technological and environmental dangers. While applied widely in the social problems and technological hazards literature, constructionism has seen little application in the natural hazards literature apart from Stallings' analysis (1995).

As disasters continually demonstrate, the material world poses very real risks that can injure or kill. However, only some of the many potential risks are ever selected to be objects of social concern, and they become concerns only if they are made salient to those 'at risk'. Constructionists focus on the social processes in which claims of danger and harm are promoted in a multiplicity of arenas from social protest movements to scientific conferences and congressional hearings (Hannigan 1995). According to more radical theorists, all aspects of the environment are constructed linguistically as social categories and the materiality of the world external to humans cannot be separated from the language used to describe it (cf. Benton 1993; Dickens 1996; Ferguson 1993). Risks associated with technological hazards are often socially constructed in the grassroots mobilization and social movement activities that coalesce around the issues (Szasz 1994). In the absence of social movements active in promoting disasters as social issues, constructionist approaches have received little attention among disaster sociologists.

Within the realm of disaster sociology, Stallings' analysis is a useful alternative to an objectivist position that privileges scientific discourse as ineluctably accurate and ignores the socially constructed nature of scientific claims and their constant revisions and contestations (Knorr Cetina 1992; Milton 1996; Wright 1995). Stallings' approach describes how US earthquake risks have been socially constructed by professional claims-makers as a problem warranting federal attention. He provides an important link between disaster sociology and the current emphasis on language and discourse that characterizes the 'linguistic turn' in the social sciences (e.g. Best and Kellner 1997). Stallings (1995) examines three basic facets in the construction of earthquake risks in the US: the claims about the risks, the claims-makers, and the claims-making process. As a sociologist he does not evaluate the scientific adequacy of earthquake risk estimates; rather he examines claims on the basis of the rhetorical devices used: How is the earthquake problem presented discursively? How do claims attempt to persuade their intended audience? What kinds of warrants are made for social action (cf. Best 1989)? Similar questions are posed about the claims-makers: Who are they? What are their professional affiliations? What are their particular interests in the successful promotion of the claims? And lastly the claims-making process itself raises such questions as: Who are the claims aimed at? Was the intended audience persuaded? Were counter-claims by competing groups made?

Stallings' work focuses on how earthquake risks have been socially constructed. Although he does not apply a constructionist analysis to an actual disaster, it is an approach that can provide insights to various aspects of disasters. An area receiving increasing attention from researchers is how disasters are socially constructed through representations in the mass media (e.g. Garner and Huff 1997). Benthall (1993: 11) observes that 'the coverage of disasters by the press and the media is so selective and arbitrary that, in an important sense, they "create" a disaster when they decide to recognize it'.

Given the media's power to establish social and political agendas by creating news stories (Herman and Chomsky 1988), a constructionist approach raises questions of what role economic and political interests play in determining what becomes news for public consumption (Gamson and Modigliani 1989; Ploughman 1997). For commercial enterprises, a primary consideration in the selection of what is 'newsworthy' is will it attract an audience. Disasters in which the audience can identify with victims warrant more coverage than those that do not, a key issue in whether First World media cover Third World disasters. Thus Benthall (1993: 27) concludes that 'disasters do not exist – except . . . for those who suffer in their aftermath – unless publicized by the media. In this sense the media actually *construct* disasters' (italics in original). At the postmodern extreme of constructionism is Baudrillard's analysis (1995) of how the media transformed the all-too-real 1991 Gulf War, in which more than 100,000 died and many more suffered vast deprivations, into a mass mediated virtual reality, a phantasmagorical 'non-war'.

A second line of constructionist investigation, as yet largely unexplored, is to examine the nature of official claims making after disaster. In the US, with its large national response infrastructure, local and state political agents must assemble claims about the extent of disaster damage and present those claims as warrants for action from the federal government. Thus, preliminary damage estimates, which are often based on only cursory information and can vary widely in the weeks following a disaster, function as instruments of political leverage for resources. In the case of Northridge, magnitude, death toll, number of injured, and total damages vary in different reports (e.g. Comerio 1996; Comerio *et al.* 1996; OES 1997; Shoaf 1997). Consensus on the 'objective facts' of disaster may take some while to attain, and many initial estimates of losses can be unreliable. Claims-making after a disaster can be seen as an arena of competition in which different organizations, munici-palities, or regions, vie for disaster resources based on socially constructed and politically promoted claims for assistance.

Not only the nature of the claims, but how they are assembled and presented, and by whom, can influence the flow of resources into an area. As Albala-Bertrand (1993) argues, governments will sometimes let early and inflated death tolls and loss figures remain as official estimates in order to gain foreign assistance, making official disaster statistics suspect at best. Thus, whether through official claims-making, or media representations, disasters can be understood as socially constructed from a multiplicity of perspectives (Ploughman 1995). Even the facts of disaster, from the number of dead to the total economic losses, often are fluid and malleable to different political ends.

Anthropological approaches

Recent anthropological research focuses on the effects of colonialism, underdevelopment, social marginalization, and environmental degradation on the consequences of disaster (Copans 1983; Johnston 1994; Oliver-Smith 1986; Zaman 1989). Investigations of how people in different cultural communities cope with environmental hazards and disasters is an outgrowth of an anthropological interest in holistic and comparative approaches (e.g. Torry 1979, 1986). Anthropological research, which often deals with peoples outside a Euro-American context, has been of relatively little interest to those who focus on US disasters. The dominant view's concerns with impersonal forces and standardized measures appears at odds with anthropological studies that emphasize ethnographic detail, local cultural diversity, individual agency, and the 'voices' of their study participants through dialogic qualitative interviews (e.g. Schulte 1991; Vaughan 1987; Wiest *et al*. 1992). On the other hand, qualitative approaches counterbalance the disembodiment, abstraction, and objectification of people and their experiences when they are reduced to 'variables' in quantitative disaster surveys.

Some anthropological studies parallel sociological research in that they focus on social and cultural responses to disaster. Several recent studies have looked at the impacts of the 1989 *Exxon Valdez* oil spill in coastal Alaska, comparing the social, cultural, and psychological effects on white and native villages (Impact Assessment Incorporated 1990; Palinkas *et al*. 1993; Picou *et al*. 1992). This research documented the vulnerability of people in subsistence economies to social and cultural disruption by a disastrous contamination event and the equally disruptive effects of large-scale, high technology clean-up efforts. In both white and native Alaskan villages, the sudden influx of several billion dollars in clean-up funding created social divisions and conflicts (Picou *et al*. 1992). Similarly, in research on the 1970 Peruvian earthquake, Dudasik (1982) documented the cultural changes that could be provoked by the influx of large amounts of disaster relief aid in regions with limited resources and a highly stratified social order.

Much of the anthropological research on disasters addresses issues of political conflict and social change. The social definition and symbolic representation of disasters is a contested political activity, and the availability of relief assistance can be directly effected by the voices represented in such cultural constructions (Loughlin 1995; Oliver-Smith 1996; Shkilnyk 1985). These constructionist contentions subvert the notion that disasters are consensus events that are objectively describable independent of particular standpoints. The politics of representation can be found in anthropological research on US disasters as well. Following the 1989 Loma Prieta earthquake in northern California, the local Mexican community organized public protests and a resistance movement to publicize and seek redress for discriminatory assistance programs and the lack of affordable housing (Bolin and Stanford 1991; Laird 1991). The protests offered a counter-image to one in which the government is portrayed as equitably distributing humanitarian assistance. The grassroots organizing drives also challenged a hegemonic Anglo political structure in the town of Watsonville. Similar political mobilization and claims-making activity was observed after the Mexico City earthquake of 1985, resulting in shifts in local political power arrangements (Poniatowska 1995; Robinson *et al*. 1986).

Issues of social power in disaster receive innovative treatment in the work of Hendrie (1997). Proceeding from a poststructuralist stance, Hendrie challenges the idea that governments and disaster relief programs are politically neutral and have unassailable humanitarian goals. Instead, she argues that disasters and relief should be seen as exercises of power and the effects of power (see also Keen 1994). By calling attention to the official discourses of disaster, the ways disasters are constructed and represented within a scientific frame, she offers insights into the creation of a technocratic and universalized management approach that abstracts the disaster from 'locally based and historically specific processes' (Hendrie 1997: 63). Hendrie's and Keen's studies draw upon a Foucauldian suspicion of modernist institutions, science, social technologies, and the power effects of official discourses and practices (e.g. Foucault 1980), applying this perspective to understand complex processes of disaster as a diversity of power relations.

Other anthropological studies of disaster have attended to issues of development, social change, and vulnerability, with an emphasis on social practices and political economic forces that shape human–environment relations in the context of underdevelopment. Oliver-Smith's (1982, 1986, 1991, 1994) long-term research on the 1970 Yungay, Peru earthquake disaster of 1970 provides perhaps the most contextualized research on a disaster available. In his analysis, the legacy of Spanish colonialism played a critical role in producing a population that is spatially, socially and economically vulnerable to environmental hazards. Colonially imposed building and settlement patterns, combined with economic policies that produced marginalization and rural poverty, laid the groundwork for the catastrophe over a period of five centuries. Oliver-Smith (1994) shows that it is was not simply bad judgement or 'bounded rationality' that led people to occupy unsafe ground; instead, the constraining effects of historical political-economic forces in people's everyday lives placed them at risk of environmental hazards.

Political economy/political ecology

Political economic theory was emphasized in the early development of a vulnerability approach, focusing attention on class, property relations, and socioeconomic inequalities in poor countries of the South (Susman *et al*. 1983; Watts 1983). The argument was put forward that it was in the social relations of production, particularly the effects of capitalist penetration and com-moditization of subsistence economies, that vulnerabilities to environmental hazards are produced (Watts 1991). Pioneering work in this area (e.g. Blaikie 1985; Blaikie and Brookfield 1987) did not focus directly on hazards but rather on more general issues of land degradation produced by political, economic, and social marginalization in the context of 'development'. In this analysis, marginalized populations are driven to exploit increasingly marginal land, degrading it further, as they attempt to survive in a wage-based economy from which they have been largely excluded. Such analysis points to the power of capital that drives the marginalization process in the South as the agents of capital produce both wealth and deepening poverty, expropriate land, degrade environments, and exploit resources for export. The next analytical step extended beyond environmental degradation and began to examine how the same processes also produce populations vulnerable to disaster (Blaikie *et al.* 1994).

Political-economic approaches to society–environment relations, now commonly labeled political ecology (Moore 1996; M. O'Connor 1994), engage in critical analyses of capitalism and its complex interactions with physical environments (Benton 1996; Harvey 1996). Recently, some political ecologists have also begun to incorporate poststructuralist and feminist theory to analyze environmental issues in relation to production and development (Rocheleau *et al.* 1996). The turn to poststructuralism, particularly in its Foucauldian version, provides a critical approach to the grand narratives of technocratic modernization that accompany 'market triumphalism' (Escobar 1996). Political ecology has expanded its theoretical base and is now in a 'dialogue with a larger intellectual environment – ideas drawn from poststructuralism, gender theory, critical theories of science, environmental history, and Marxist political economy . . . ' (Peet and Watts 1996: 9). Contemporary political ecology applies critical theory to analyze the society–economy–environment nexus, particularly as it relates to developmental processes and environmental degradation (Benton 1996; J. O'Connor 1994). Vulnerability analysis, with its attention to political and economic forces that shape inequalities and local practices and their implications in disaster causation, is convergent with political ecological approaches to environmental issues.

Famine studies

It is in famine studies, conducted primarily by social geographers and anthropologists, that we see the first conceptualizations of disasters as social, political, and economic processes related to dynamics of marginalization.[6] Political economic analysis of famines (e.g. Downs *et al.* 1991) shifted the focus from droughts as causes of famines to social and political processes that limit people's abilities to avoid hunger, destitution, and starvation (e.g. Sen 1981; Wijkman and Timberlake 1988). In famines, environmental factors (droughts, floods, bio-invasions) are often insignificant in comparison to the larger political and economic processes leading to loss of food access (Watts 1991: 23). One element from the research that has become important in vulnerability analysis is 'access to resources'. This concept saw expression in Sen's (1981; Dreze and Sen 1989, 1990) use of the term 'entitlements' in the social explanation of famines. An entitlement is a person's right of access to resources through ownership, wage labor, or social exchange.[7] Famines occur when food entitlements decline, that is when people's rights of access to food become impaired by a variety of social and economic conditions. If, for example wages fall below a subsistence level and a person lacks other food entitlements through kin groups, common property resources, or social exchange relationships, entitlement failure can lead to malnutrition and ultimately starvation. According to Sen's approach, famines are not necessarily caused by an absolute decline in food availability. Rather, they are a result of declines in people's access to food and the unequal distribution of food entitlements by class, ethnicity, gender, and age (see also Shipton 1990; Watts 1991).

While famines have accounted for far more deaths in the twentieth century than other 'natural' disasters, they have occurred primarily in countries of the South (including China) (Downs et al. 1991), and seem to hold little interest for Northern disaster sociologists. However, there is considerable theoretical innovation in the famine literature that can be profitably explored. The empirically rich literature on famines provides insights into the complex social causation of such calamities and illustrates the ways burdens can fall most brutally on women, children, and subaltern groups (Ali 1987; O'Brien and Gruenbaum 1991; Sen 1988; Vaughan 1987). Famines, through hunger, starvation, and displacement, lay bare the unequal distributions of power, privilege, and hence security, in a given society (Hendrie 1997; Schroeder and Watts 1991; Shipton 1990).

Famine studies reveal how the penetration of capitalism into rural agricultural societies and the creation of wage labor opportunities for some segments of a population have resulted in sharp increases in socioeconomic inequality, particularly gender-based inequality (Shiva 1989). That inequality is tightly coupled with differential vulnerability to various environmental hazards (Blaikie et al. 1994). The poor can sometimes no more afford food than they can afford secure housing. However, simply to focus on the global forces of capitalism would be to ignore the strategies that subaltern groups use to resist or cope with those forces (Scott 1990; Watts 1991; Woods 1995). In the case of famines, victims' agency is expressed in the form of coping strategies covering a wide range of actions including harvesting 'famine foods', asset stripping, seeking wage labor opportunities in urban areas, and out-migration. Strategies often differ by gender and age within households, as well as by the cultural repertoires of different classes and ethnic groups (Brown 1991; Goheen 1991; Sen 1988; Vaughan 1987). Famine studies sensitize investigators to the different concatenations of environmental, social, cultural, political, and economic factors that come into play at given times and places (Copans 1983).

Vulnerability analysis

Vulnerability analysis examines how social inequalities are structured by the social relations of production and how those inequalities affect people's abilities to cope with environmental risks (Watts 1991). There have been several schematic statements of the major elements and internal relations of vulnerability in recent publications (Blaikie *et al.* 1994; Cannon 1994; Hewitt 1997; Wisner 1993). While Blaikie *et al.*'s formulations are the most ambitious, all share the foundational notion that risks are 'a complex combination of vulnerability and hazard' (Blaikie *et al.* 1994: 21). Vulnerability analysis incorporates understandings of physical hazards, the various ways people are subject to their effects, and how risks are produced and vary over time. Vulnerability is seen as an ongoing product of social, political, and economic factors that create differences in people's susceptibility to harm from environmental hazards. This contrasts with more common usages where 'vulnerability' is used to refer to the characteristics of the built environment, entire urban areas, or even 'social systems' (e.g. Pelanda 1982). As Hewitt writes (1997: 11) 'The origins of disasters and the patterns of impact invariably reach back into the general contexts of material life, its environmental relations and geographical settings.' Thus, it is necessary to move beyond looking at disasters as simply physical events and consider the social and economic factors that make people and their living conditions unsafe or secure to begin with. Fragile livelihoods are as important as fragile buildings in understanding vulnerability to environmental hazards. The value of this approach is underscored by research that demonstrates how large-scale technocratic 'modernization' projects often drive the poor onto marginal lands and disrupt people's livelihoods while benefiting only elite segments of a society (Blaikie and Brookfield 1987; Booth 1995; Escobar 1996). The research counters claims that development and modernization will necessarily reduce the risks of disaster (cf. Burton *et al.* 1993).

By defining vulnerability in terms of people's capacity to avoid, cope with, and recover from the effects of environmental hazards, attention is drawn to their living conditions, social and economic resources, livelihood patterns, and social power. While exposure to risk is a necessary element in vulnerability, it is people's lack of capacities – a lack which is socially produced – that turns an environmental hazard into a disaster. People's capacities relate to the kinds of resources they can mobilize in the face of calamities. Community preparedness and the abilities of political structures to organize and provide resources are components of local capacities to deal with hazardous environments. Vulnerability analysis gains complexity in seeking to analyze the multiple and intersecting ways that risk is produced and reproduced under varying political and economic conditions. Because much of the work has concerned countries of the Third World, the concept is often linked to poverty and the developmental processes that produce deepening poverty for many in a given country (Anderson and Woodrow 1989; Cuny 1983). Poverty, structured by class, race, ethnicity, and gender, is seen as the particular expression of more general political and economic processes that unequally allocate resources and risks.

Blaikie *et al.*'s (1994: 24–25) general approach maintains that vulnerability is traceable to three linked realms: root causes, dynamic pressures, and unsafe conditions. The 'root' or underlying causes refer to the general historical, political, economic, and demographic factors that produce unequal distributions of resources among people, often according to gender, race, class, ethnicity, and age. These root causes are translated into specific kinds of vulnerability through a number of different 'dynamic pressures' or social processes, including, for example, rapid urbanization, environmental degradation, economic crises, political conflict, declining wages, and structural adjustment programs (Blaikie *et al.* 1994). These processes produce vulnerability by creating 'unsafe conditions' under which some in a given place must live. And while unsafe conditions may involve both spatial location and characteristics of the built environment, they also include fragile livelihoods, physical susceptibility through illness, inadequate incomes, legal and political inequities, and a lack of social preparedness for emergencies (Wisner 1993). People's lack of social power – their inability to influence the conditions under which they live and their inability to affect larger societal matters – is a central issue in who is vulnerable (Hewitt 1997: 151).

Focus on the specific components of vulnerability requires an examination of people's coping capacities in relation to resource availability. This is the nexus where the constraining and enabling effects of social structures express themselves in agency – the extent to which people can strategically control their own lives (cf. Giddens 1984). Resources can involve a variety of material and social assets including finances, information, social support networks, employment and other income opportunities, legal rights, and political power. Access to resources is always differentially allocated and can bear directly both on the kinds of losses people experience in disaster and the recovery strategies they employ. Resource allocation can change as a result of disasters – particularly large and destructive disasters – both because livelihood patterns are altered and because new resources may become available in the aftermath.

Access to resources is central to Wisner's analysis (1993: 21), which identifies three major components of vulnerability. The first is *livelihood vulnerability* or people's abilities to gain resources through employment, subsistence activities, or related means. Those with higher incomes and secure employment are said to have resilient, as opposed to vulnerable, livelihoods. People's abilities to resume their livelihoods after a disaster is key and varies considerably with the type of production regime and the social relations of production. A landless peasant may be plunged into destitution after a flood due to the loss of income earning opportunities while the peasant's landowning employer, on the other hand, can live off accumulated assets until the land is again ready for another cycle of crop production (Winchester 1992). Livelihood vulnerability varies by social class, gender, race/ethnicity, age, and the nature of state actions to promote or restrict livelihood opportunities.

The two other components of vulnerability are *self-protection* and *social protection* (Wisner 1993: 19–21). These pertain to issues of preparedness and hazard mitigation at both the household and larger (community, region, society) spatial levels. Self-protection involves people's physical exposure including *where* and *in what* people live, and *when* (time of day, season) they occupy structures for work or living. Self-protection directly relates to socio-economic factors (income in particular), cultural traditions and preferences, as well as local technical knowledge about how and where to build. In contrast, social protection concerns the organizational and regulatory realm. Technical knowledge of hazards, building code regulations, and land use planning can all be considered parts of social protection. State and local level organizations designed to manage emergencies and assist in disaster response likewise can be instrumental in social protection. Hewitt (1997: 188) maintains that the key element in social protection is a commitment to equitable protection, not the level of technological sophistication brought to bear by the state or its agents. Indeed, as Wisner contends, the state can be in a contradictory position of offering 'social protection to reduce vulnerability . . . [while] it is normally a party to the economic and social processes that lead people to be unable to protect themselves in the first place' (1993: 21). In the North American context, the state (at various political levels) is in a contradictory position of being a major advocate of urban growth, attracting people to hazardous locations, while at the same time being responsible for managing the disasters that growth policies make likely.

VULNERABILITY IN THE FIRST WORLD

Conceptually, vulnerability analysis has affinities with US research on structured social inequalities (e.g. Morrow and Enarson 1996; Phillips 1993). US researchers have for some time recognized that there are 'at risk' groups in disasters and these have been variously identified as including low income households, ethnic minorities, elders, and female-headed households (Bolin 1994a; Peacock *et al.* 1998). On our reading, the advantage in a vulnerability approach is that it is both context sensitive and historically informed, going beyond simply categorizing groups at risk. It promotes a fuller understanding of the production of vulnerability through specific economic and political factors. While it is valuable to identify vulnerable groups through post-impact studies (e.g. Bolin and Bolton 1986), a more complex understanding is gained by examining the historical factors that produce and reproduce vulnerability in given locales. This in turn has broader policy implications than does an approach that only focuses on returning the community to 'normal' through repairing what the hazard agent damaged. By searching for the *social causes* of disaster, vulnerability analysis adds depth to the already well-documented study of the *social effects* of disaster.

A starting point for the application of vulnerability analysis to industrialized countries is a consideration of differences between rich and poor countries (cf. Wisner 1993). In 'wealthy' countries such as the US – with an elaborate, well-organized, and well-funded disaster response system, and with the availability of property insurance and financial credit – many are able to avoid or cope with disasters in ways not available in much of the world. Yet aggregate wealth says little about the distribution of that wealth, particularly in situations where economic restructuring is producing intensified income polarization (Faux 1997). Vast asymmetries in resource access and assets are found virtually everywhere, and such asymmetries are becoming intensified in countries where neoliberal economic policies are enriching the already wealthy while initiating draconian reductions in welfare entitlements for the poor (e.g. Harvey 1996; Wolch 1996). While the extent and qualitative nature of vulnerability may be different between North and South, environmental risks and the effects of disaster in the North can be unevenly distributed by class, race, ethnicity, gender, and age (Bolin and Stanford 1991; Morrow and Enarson 1996; Peacock *et al.* 1998; Wisner 1993).

In both the First and Third Worlds marginalization processes create vulnerability by placing certain segments of a society in a precarious position between the inequitable conditions of the political and economic systems, on the one hand, and the risks of the environment, on the other (Cannon 1994: 15). The poor often live in less safe structures – from a mobile home in Florida to an adobe house in the highlands of Guatemala – that are more likely to be damaged or destroyed in disasters. However, the spatial and physical aspects of vulnerability tend to be much more pronounced in the Third World, where the poor are often forced to live and work in persistently hazardous areas (Hewitt 1997). In contrast, socially and economically marginalized populations in countries such as the US are not necessarily driven to live in areas at greatest risk from natural hazards, however much they may be otherwise spatially segregated by race and class. Indeed, the wealthy may voluntarily choose to live in physically hazardous settings, such as in 'view homes' perched precariously on the wildfire and earthquake ravaged hillsides of California or a beach-side mansion in Florida awaiting the next tidal surge from a hurricane. Location is important primarily in the obvious sense that to live by a river means one is subject to a potential for floods and to live in 'earthquake country' means one is exposed to the possibility of earthquakes.

Vulnerability in the First World is revealed most clearly in people's inabilities to cope with losses and their lack of access to recovery assistance, elements usually associated with class, ethnicity, and/or gender (Bolin and

Stanford, 1991; Fothergill 1996). That is, while the wealthy certainly may suffer losses in a hazard event, their property insurance, assets, financial credit, and stable employment generally secure them against the destitution that can befall a poor family exposed to the same event. Further, while the absolute amount (in dollar terms) of disaster-related losses are often much greater for the middle class and the wealthy than for the poor, in relative terms the poor typically lose a larger part of their material assets and suffer more lasting negative effects of the disaster (Bolin 1994a). Vulnerability goes well beyond physical exposure to hazards, encompassing social disadvantages, a lack of response capacities, and powerlessness (Hewitt 1997). The technocratic approach that invests heavily in engineered hazard abatement and geohazard mapping tends to see risk solely in terms of location-based physical exposure, neglecting the underlying social and economic dimensions that compound the risks people face. Cannon (1994) refers to this as a package or 'bundle' of factors that together hinder a person's or household's ability to cope with a disaster. The bundle represents a compilation of social, cultural, and economic characteristics that cumulatively disadvantage people and produce vulnerable households.

In the case of technological hazards in the First World, the relationship of class and race to physical exposure is much more clearly drawn than for natural hazards. In North America it has been shown that poor and ethnic minority communities (including native Indian reservations) are more likely to have toxic waste sites and hazardous chemical plants nearby as they lack the political power to resist siting decisions (Bullard 1990; Johnston 1994; Shkilnyk 1985).[8] Corporations often target such communities specifically because the assumption is made (often incorrectly) that the poor will welcome the hazardous facilities as sources of employment (Szasz 1994). Low income and ethnic minority communities in the US are also less likely to be targeted for environmental remediation by federal toxic waste clean-up and pollution control programs than are wealthier communities (Schnaiberg and Gould 1994).

The state role in social protection is a significant element in vulnerability and disaster response in both North and South. Researchers studying Third World disasters question the notion of the state as a positive factor in disaster response (Wisner 1993). Given the many cases of politically created famines, exposure of populations to pervasive hazards, discriminatory practices in targeting recovery assistance, and state promotion of dominant class interests, a suspicious stance toward state power seems warranted (Hendrie 1997; Oliver-Smith 1986; Waterstone 1992). Third World research indicates that economic elites often capture the relief assistance offered by the state and NGOs and that relief aid generally reinforces the status quo of social inequalities (Maskrey 1994).

There are clearly disagreements over the role of state, ranging from suggestions that state planning can alleviate vulnerability problems (Burton *et al.* 1993) to allegations of the state's frequent participation in creating vulnerable populations (Hewitt 1983b). This suggests that it is important to interrogate the state's role both in the creation of household vulnerability and in assisting vulnerable households following natural disasters. In the First World, the state may take an active stance in assisting victims after disasters, although the level of state assistance can vary widely (cf. Bolin 1994b; Hirose 1992). Disaster relief is not necessarily apolitical nor is it necessarily equally beneficial to all needing it. Racial and ethnic discrimination in relief has been documented in recent US disasters, and a persistent class preference exists in the federal programs privileging middle-class homeowners over renters and the homeless (e.g. Comerio *et al.* 1996; Schulte 1991; Wallrich 1996). Still, the overall role of the state in social protection and disaster response in the US has been extensive and, as of late, reasonably effective in short-term assistance.

A framework of First World vulnerability

Based on both existing 'models' of vulnerability and factors that appear important in the US and other industrialized countries, we offer an outline of a theory of vulnerability, identifying factors that should be considered in discussions of First World disasters (Figure 2.1). We stress that these are not discrete categories, and taken individually are not sufficient to identify vulnerable people. Nor are all factors necessarily relevant in a given context. Rather the elements must be considered in terms of how they combine in different settings, and taken together how they act to inhibit people's abilities to avoid disaster or to acquire the resources necessary to cope with a disaster's effects.

Social class

As indicated in a multiplicity of studies, class position is a central component of vulnerability, particularly in combination with other marginalization factors (Blaikie *et al.* 1994). It is a general indicator of access to resources that includes employment (type and stability), income, savings, and education levels. Low income households may lack the means to prepare for hazardous environments and likewise lack the necessary resources to cope with the effects of disaster. Income can clearly influence a household's ability to make necessary repairs and reconstruct heavily damaged residences. However, as cautioned above, while income and other class characteristics may indicate vulnerability, this does not mean that those with the fewest assets will necessarily have the most difficulties in coping with calamity. Income provides a general indicator of vulnerability by marking categories of households, but the 'bundling' of other vulnerability factors can add complexity to simple categorization.

An important aspect of class is people's residential tenure. Whether people own or rent their homes can influence their ability to respond to calamities. Those who rent are subject to the whims of landlords regarding cost, security, and safety standards. After a disaster in the US, renters are eligible for different types and amounts of government assistance than are homeowners. Homeowners generally get more generous assistance, but in turn their liabilities are much higher. Homeowners, for example, may have to obtain a loan to repair a damaged home, thus incurring long-term indebtedness (Bolin 1994b). Renters on the other hand are more easily displaced and with current real estate inflation in some urban areas, they may have trouble finding affordable replacement housing.

A related class-based phenomenon, particularly in the US, is the increase in the number of homeless persons (e.g. Phillips 1993). Homelessness and those housed in provisional shelters, are indicators of the changing nature of class-based entitlements in the US. It is a phenomenon that has grown steadily in the US over the last two decades, particularly in urban areas (Wolch 1995). Recent estimates place Los Angeles' homeless population at between 40,000 and 70,000 per night (Soja 1989; Wolch 1996), indicating a new dimension of class-based vulnerability produced by economic restructuring, unregulated markets, and sharp cuts in social spending for affordable housing.

Gender

Until recently gender has been largely ignored in the First World literature on disasters outside of mental health studies (Fothergill 1996; Morrow and Enarson 1996). In the US literature issues of gender *inequality* are usually obscured by limited discussions of gender *difference* (Bolin *et al.*, in press). That is, rather than examine gender as a hierarchical system of unequal access to social and economic resources, studies generally treat gender as a 'background characteristic' that may be statistically associated with some dependent variables. For vulnerability analysis, gender is understood not as an abstract bivariate category, but as a feature of people's social existence that structures their lives in a multiplicity of ways. It is well established that women throughout the world earn less than men and generally have fewer assets and resources. While there have been major increases in US women's labor force participation over the last two decades, women generally have significantly lower incomes than men and there have been steady increases in the incidence of poverty among female-headed households (Gimenez 1995). Whether the feminization of poverty makes such households more vulnerable to disaster and under what conditions is only starting to be investigated (Morrow and Enarson 1996). Gender effects cannot be separated from class, ethnicity, and age, as multiple and combining factors that may significantly disadvantage some households over others in disaster (Enarson and Morrow 1998). Finally, gender-based inequities in power and resource distribution can characterize intrahousehold relations (Wolf 1991), raising questions about household unity and cooperation in calamities, questions that are as yet unexplored in First World contexts (cf. Shiva 1989).

Race/ethnicity

In the polyethnic societies of the First World, race/ethnicity may be linked to differences in losses experienced and the ability to acquire resources after disaster. Ethnic variations in resource access have been documented in numerous US disasters (Bolin and Bolton 1986; Bolin and Stanford 1991; Phillips 1991; Schulte 1991) and receive extensive discussion in recent work on Hurricane Andrew (Peacock *et al.* 1998). It is important to analyze the ways that race or ethnicity as social identifiers are associated with marginalization processes in specific contexts. As in the case of gender, class factors are imbricated in the salience of ethnicity to hazard vulnerability. In the US, African-Americans, most Latin American ethnicities, and a number of South Asian immigrants are over-represented among those living in poverty (Ong and Blumenberg 1996; US Bureau of the Census 1992). Yet the majority in each of those groups do not live below the poverty line, suggesting caution in considering ethnicity in isolation from other social characteristics. If ethnic groups are largely isolated socially, economically marginalized, and persistently discriminated against in a given community, then ethnicity can be a significant factor in vulnerability (Lamphere 1992; Peacock *et al.* 1998). Consideration must be given to how ethnicity is socially constructed, how it relates to socioeconomic factors and immigration status, and how those may translate into various risk factors.

Age/life cycle

A person's age or the age composition of a household has implications for the risks they may face from environmental hazards. Researchers cite examples of elderly households that are physically vulnerable since members lack the mobility and physical ability to avoid disasters, cope with crises, or acquire recovery resources (e.g. Aptekar 1994; Bolin and Klenow 1988; Eldar 1992). In some instances elders are over-represented among the casualties of disaster due to frailty and being housebound (IFRCRCS 1996; Hewitt 1997). Obviously age is not a fixed category and the age composition of households changes continuously. Social class is implicated here as many older households with retired wage earners have reduced incomes, live on pensions and have limited access to other resources (Phillips 1993). Having limited fixed incomes can also restrict access to credit and loans, and consequently some households may be 'squeezed' out of an area because they lack access to the resources necessary to rebuild after a disaster. Neoliberal policies in some First

World countries have led to cuts in entitlement programs for older persons, including health care benefits and cost of living adjustments in pensions, further straining lower income elder households.

Migration/residency

Increases in international immigration over the last thirty years have produced significant demographic and urban transformations in the First World (e.g. Sassen 1988, 1992; Rocco 1996). The Los Angeles area is perhaps exceptional in that it has been the destination of choice for millions of Asian and Latino immigrants over the last two decades producing a complex ethnic landscape. Among immigrant groups, there is much diversity and social division based on ethnicity, class, language skills, and gender (Lamphere 1992). Recent immigrants have experienced a significant rupture with a former way of life and must re-learn how to negotiate their way through everyday life in their new community (Rocco 1996). Given the cultural diversity of immigrants in the US, combined with the variety of legal statuses, class, and employment positions, no single characterization can adequately grasp this complexity. A given household may be split by residency differences and citizenship rights, depending on when one particular household member entered the US and under what conditions. It cannot be assumed that households constitute 'units' in which all members have equal rights and resources (cf. Sen 1988). Recent immigrants who are poor often live in marginal and crowded housing conditions, particularly in cities with high housing costs such as Los Angeles and New York City.

Language/literacy

While this category may overlap with ethnicity and immigration, we use it to identify a dimension that has particular relevance for vulnerability analysis; specifically when language and literacy capabilities can present barriers to information and resources in disaster (Subervi-Velez *et al*. 1992). As a result of changing immigration patterns, the US has a growing diversity of cultures and language groups, including areas where there are significant numbers of non-English speakers. The salience of English competence in the US has become intensified with shifts in political culture and efforts at marginalizing immigrants. The last two decades have seen a militant nativism that has placed 'English only' ordinances in numerous state and city legal codes in an effort to exclude non-English speakers from full political participation (e.g. Horton 1992). Language skills, literacy, and educational levels are often closely related, for native populations and immigrants alike, adding a social class dimension. Literacy and language skills can be important in determining a person's ability at negotiating the complex, and sometimes arcane, procedures to obtain federal and state assistance. Those familiar with the bureaucratic culture of the First World may be significantly advantaged over recent immigrants unfamiliar with such procedures, and those differences can affect acquisition of needed resources (Schulte 1991).

Political culture and social protection

The nature of political ideologies and practices by the state can produce varying levels of social protection (Cannon 1994). Social protection from hazards can be gained through land use planning, building codes, emergency preparedness, trained disaster response organizations, and recovery programs. On most counts, social protection against disaster is more developed in the North than South, but it is far from uniform within either. The technologies adopted in the cause of social protection, such as hazard mapping, storm detection, warning systems, and earthquake resistant construction, are expensive and unevenly applied. These technical aspects of social protection require economic surpluses that can be devoted to such practices. Equally important it requires the political commitment to appropriate part of the private surplus to social ends in the protection of civil society. Social protection against hazards may be promoted through a complex of local, provincial, and national governmental and non-governmental organizations involved in emergency preparedness and disaster response. In the US, expenditures through FEMA alone are typically measured in billions of dollars per year, reflecting both the property losses of major disasters and the scope of programs offered to victims (e.g. FEMA 1995). The US federal government is currently a major buffer of the effects of disasters on citizens, although the access to relief has not been uniform across class and ethnic groups (Dash et al. 1998).

While the state in the First World may offer significant resources to promote public and private sector recovery after a disaster, the state can also contribute to political vulnerability by intentionally disadvantaging certain groups (e.g. Hewitt 1997). Recent changes in FEMA regulations imposed in congressional budget appropriations deny all federal assistance and resources to those who cannot prove they are legal residents in the US (Wallrich 1996). Applicant eligibility based on claims of residency are subject to checking by the Immigration and Naturalization Service and those entering false claims are subject to deportation. The dismantling of welfare programs in the US is also likely to increase the vulnerability of marginal populations by politically denying them needed support. Given the panoptic view of the population afforded by federal computer networks and geographic information systems (GIS) technologies, the potential for both humanitarian action and politically inspired surveillance coexist. Nor is the US and other national governments beyond suspicion in the exposure of citizens to toxic hazards, hazards that they are the sole regulators of (Bullard 1990; Gould et al. 1990; Wasserman and Solomon 1982). The actions of the state in providing social protection

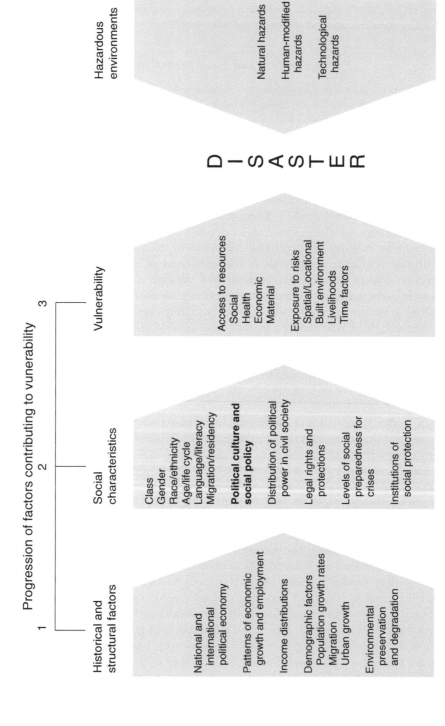

Progression of factors contributing to vunerability

1   2   3

Historical and structural factors

National and international political economy

Patterns of economic growth and employment

Income distributions

Demographic factors
Population growth rates
Migration
Urban growth

Environmental preservation and degradation

Social characteristics

Class
Gender
Race/ethnicity
Age/life cycle
Language/literacy
Migration/residency

Political culture and social policy

Distribution of political power in civil society

Legal rights and protections

Levels of social preparedness for crises

Institutions of social protection

Vulnerability

Access to resources
Social
Health
Economic
Material

Exposure to risks
Spatial/Locational
Built environment
Livelihoods
Time factors

Hazardous environments

Natural hazards

Human-modified hazards

Technological hazards

D
I
S
A
S
T
E
R

Figure 2.1 Progression of factors contributing to vulnerability

against environmental hazards must be examined in terms of its effects, not in terms of intentions, goals, or political ideologies (e.g. Hendrie 1997).

Most of the factors contributing to vulnerability in the First World that we have outlined here are quite similar to those in Third World contexts: class, gender, race/ethnicity, and age/life cycle. In addition, immigrant status and literacy can figure prominently in some First World disaster settings. State-sponsored social protection programs generally play a larger role in mitigating the effects of disaster in the First than the Third World. These various dimensions affect households' and individuals' legal entitlements and are linked to their abilities to acquire resources. No characteristic is intrinsically a factor in the production of vulnerability to environmental risks nor will all necessarily be important in a given disaster. The factors acquire salience in the different histories of places in combination with ongoing features of political culture and social policies. The recent abrasive political discourse that has emerged in various US states (e.g. California, Florida, and Texas) over 'illegal immigrants' creates a political vulnerability for legal immigrants who avoid federal programs they are entitled to use for fear of jeopardizing their residency status (Collins 1997). Attention should be given to how these elements combine, producing changing vulnerability patterns with political and economic conditions locally, nationally, and internationally (Hewitt 1995).

Vulnerability is produced through people's unequal exposure risk and their unequal access to resources (Figure 2.1). Thus vulnerability analysis needs to consider both *risks* and *resources* as part of a person's overall vulnerability profile (Cannon 1994). While marginalized people in the First World may not be forced to live in obviously hazardous areas, where and in what people live nevertheless can be an important feature of their exposure to losses. Location can also be important to the extent that a given disaster may impair or destroy someone's source of employment or livelihood. Economic resources, including income, savings, credit, investments, and property, all contribute to a person's relative security or vulnerability. Social resources include the types of social support available to individuals both through local institutions and through more diffuse social networks or relatives and friends. While health resources (or their absence) may be more critical in poorer countries, the US, unlike most industrialized countries, lacks any national health care insurance and has notoriously expensive private health care. For poor persons who lack health insurance, injury or illness can result in a crushing health care burden as well as the loss of employment. While there have been no recent mass casualty disasters in the US, the limited access to health care for lower income persons is a persistent risk factor.

The elements identified in Figure 2.1 are understood to combine in different ways in actual disasters to produce secure or vulnerable people. The framework calls attention to historical conditions and current political and economic factors that shape the importance of different social elements that become part of a person's vulnerability profile. It is not presented as a causal model for quantitative 'testing' or a predictive device, but rather as a conceptual framework to illustrate relationships among categories of factors on different spatial and temporal scales. It points to both political culture and social policies that affect vulnerability through social protection activities. Political culture, as noted, can intensify marginalization through the 'scapegoating' of specific groups or by denying those groups public resources. Hewitt (1997: 154) uses the term 'positive disadvantaging' to refer to government actions and social practices that take away people's options and increase the risks they face. But the state can also promote social policies that assist people in meeting their needs after a disaster or preparing for future ones. Disasters are schematically represented (on the right side of Figure 2.1) as a process that begins at the intersection of socially produced vulnerabilities with hazardous forces in the environment. Although human impact on the environment is not included in the diagram, we recognize that human activities, in numerous ways, can and do continuously alter environments changing the physical nature of hazards in a locale.

In using this general framework we draw from both critical political economic theory and general notions of structure and agency (cf. Wisner 1993: 130). From the first, we consider the various ways different groups of people are either included in or marginalized by systems of production and the social relations of production (e.g. historical patterns of ownership, wage structures, income opportunities, unemployment, gender discrimination, land tenure systems). From the second we examine the micro-logical dimensions of how people, diversified by age, ethnicity, social class, gender, immigrant status, and so on, are able to manage their daily lives, including their exposure to hazards and their abilities to acquire resources necessary for living. Access to material, financial, and social resources becomes a critical part of coping with disaster and is the conjuncture where vulnerability can be reduced in the aftermath (Chapters 5 and 6).

VULNERABILITY AND EARTHQUAKES

We conclude our discussion by reviewing studies of several recent seismic disasters that illustrate elements of a vulnerability approach. Key features of this research include consideration of historical and contextual factors that produce differential vulnerability, analysis of the political and economic forces that force people into unsafe environments, and recognition of the diversity of local coping strategies used in disaster. We focus on earthquakes not only because they are the subject of our research, but also because earthquakes, perhaps more than any other natural hazard, have attracted a highly technocratic approach to disaster mitigation focusing on engineering solutions in the built environment. Vulnerability research suggests that a comprehensive approach will go well beyond building safer buildings. Understanding who lives in unsafe structures, why they do, and what options they have in the face of hazardous environments is fundamental.

Maskrey's (1994) discussion of the 1990 Alto Mayo, Peru earthquake illustrates the importance of charting the historical context of a region together with an examination of national and international political economy. The Alto Mayo region, located in the Amazon lowlands of northeast Peru, has a developmental and ecological history that reflects the legacy of colonialism (e.g. Oliver-Smith 1986). The area's more recent history was one of dramatic economic and ecological change. Until 1974 the area had a dual economy with production for export as well as utilization of local resources for local consumption. After a road was completed in 1974 connecting the area to western Peru and its seaports, it shifted to an almost exclusive focus on export agriculture. Natural resources were degraded in the interests of rice production, and people became increasingly dependent on cash to buy commodities they had previously produced themselves. As is often the case in an export regime, social inequalities grew and those without wage labor opportunities suffered hunger and malnutrition (Maskrey 1994: 113; Watts 1991). Due to Peru's large debt crisis and hyper-inflation, the coastal road fell into disrepair, export production could no longer be moved to the coast, and the Alto Mayo region fell into a deepening crisis. Small farmers were driven into the production of coca, which speeded the rate of deforestation as trees were felled to make room for crops.

Accompanying deteriorating social conditions, Maskrey notes that local autonomous political organizations were increasingly established in the absence of state authority. Local enterprises to address local markets were also begun to cope with the high price of commodities in the rest of Peru. These somewhat contradictory developments – social and ecological crises and grassroots community-based organizing – were playing themselves out when a magnitude 5.8 earthquake struck the region in 1990. Due to the preponderance of poorly maintained adobe and rammed earth housing (Maskrey 1994: 115), the moderate temblor collapsed 3,000 houses and killed 26 people. Ninety per cent of the housing stock in this remote area was destroyed in the main shock. As Maskrey notes, it was not the unsafe housing *per se*, but a much wider pattern of social vulnerability and ongoing political and economic crises that produced the disaster.

Maskrey criticizes the disaster response paradigm followed by the central government and national and international NGOs. For example, Alto Mayo already had surplus food rotting in warehouses due to the lack of transportation links with the coast; thus the provision of more food supplies into the area was not a sensible use of resources (Maskrey 1994: 116). The region was economically vulnerable precisely because it could not export the food it already was producing. Disaster assistance provided much-needed medicine and building materials, but these needs pre-dated the earthquake and were part of the larger pattern of marginalization in the area. On Maskrey's accounting, local organizational responses to the disaster were more effective, with area 'defense fronts' conducting needs assessments and damage surveys. The earthquake disaster was 'only another additional factor in an ongoing economic and political disaster . . . ' (Maskrey 1994: 117). While community-based organizations proved effective in the short term, they were pushed to the periphery by state responses and a plethora of NGOs that flocked into the region to offer assistance. Once the state agency left the area, most of the NGOs also left, and the residents of Alto Mayo were left to cope largely on their own.

Maskrey (1994) describes at least one effective recovery program, which was developed locally with the assistance of a Lima-based NGO with expertise in building techniques. Maskrey argues that key to this program's success was the use of local knowledge, careful assessment of local resources, and consideration of people's judgements of their own needs. Once the locally developed reconstruction plan gained legitimacy it also attracted state support and United Nations funding. This grassroots, 'bottom-up' approach contradicted the normal state-centered response paradigm of viewing people as helpless, needing professional direction and assistance. But it required the building of organizational links and middle level structures to assemble a regional approach in an area that previously lacked any political coherence. Maskrey stresses the need to consider the particularities of local variations in 'culture, ethnicity, gender, roles and technology, and . . . the implosion and fragmentation of these variables through . . . migration, communication and technological innovations' (Maskrey 1994: 122).

Both the 1976 Guatemala and the 1985 Mexico City earthquakes further illustrate social factors in the causation of disaster. The Guatemala earthquake offers one of the clearest examples of the way disasters can follow class divisions. As a result of the earthquake, 22,000 died and one million were homeless, nearly one-sixth of the country's population (Wijkman and Timberlake 1988: 90). The disaster has been called a 'class quake' due to its differential impact on areas occupied by different social classes; most of the 1,200 dead and 90,000 homeless in Guatemala City were the urban poor living in landslide-prone squatter settlements (Blaikie *et al.* 1994). Social class factors intersected with ethnicity in the post-earthquake period as both poor rural Mayan Indians and urban squatters were further marginalized by their very limited access to relief and technical support (Bates and Peacock 1987).

The underlying causes of the Guatemala disaster were chronic poverty amongst the Indians of the rural highlands and the rapidly growing urban poor population living in makeshift housing in hazardous areas on hillsides and ravines. These factors, combined with a violently repressive political regime and ongoing indigenous resistance movements, insured that relief aid and reconstruction programs became highly politicized from the outset.

Groups that were politically marginalized before the disaster sometimes experienced further political repression during relief and reconstruction. The politicization of relief assistance is not unusual after disasters in the Third World (e.g. Hendrie 1997). However, the Guatemala earthquake raised important issues about political vulnerability as a result of relief and reconstruction programs (e.g. Pettiford 1995). NGOs active in rural reconstruction encouraged local community organizing to rebuild villages. This, in turn, led to increased repression and political violence directed at Mayan villages viewed by the military as political threats because of their grassroots organizing (Davis and Hodson 1982). As indigenous communities organized to rebuild, the very act of organizing resulted in community leaders being targeted for assassination from military death squads (Anderson and Woodrow 1989). Political vulnerability after disaster, because of tribal, ethnic, or political affiliation, may be a significant element in a person's or group's overall vulnerability profile when repressive regimes hold sway.

The Mexico City earthquake of 1985, striking one of the largest urban centers in the world, illustrates the complex of historical, cultural, and political-economic factors that can concatenate to produce disaster. It also reveals political and social changes engendered by disaster as neighborhoods organized in response to existing grievances. In understanding vulnerability of the Mexico city region, Tobriner (1988) begins with the location of the original Aztec settlement six centuries ago, followed by Spanish conquest and the environmental alterations that followed. The consequence of this complex history is that today the urban core of Mexico City is built on the extremely unstable soils of a drained lake bed in an area of high seismicity and volcanism, coupled with severe air pollution problems and a rapidly declining water table (Hewitt 1997). A second level of vulnerability was produced by the types and quality of the structures built on these unstable soils (Stolarski 1991). Residential structures of 6–15 storys were particularly sensitive to the ground motions produced by the back to back earthquakes on 19 and 20 September 1985. Dozens of large housing complexes collapsed catastrophically, resulting in more than 10,000 deaths and 30,000 housing units lost (Stolarski 1991; Tobriner 1988), in spite of the epicenter being 370 kilometers to the south on the Pacific coast of Mexico.

The political and economic context of the Mexico City earthquake is central to understanding the contours of the disaster. Since the early 1970s Mexico had been undergoing a series of political and economic shocks including a growing debt crisis, internationally imposed austerity programs, and declining standards of living for most social classes (Camacho de Schmidt and Schmidt 1995). In Mexico City, demands for public services were far outstripping the ability of the public sector to provide them. At the time of the earthquake there was severely inadequate housing for the urban working classes and the poor, and many had a precarious existence in informally arranged housing (Poniatowska 1995). As Camacho de Schmidt and Schmidt

(1995: xv) observe, for many in Mexico City '"normality" had already acquired the traits of a vast disaster'. In response to government failures in the city, there were a grassroots political mobilizations beginning in the 1970s. These autonomous movements, numbering more than 180,000 participants, demanded better housing, transportation, urban services, and rights to self-governance. The movements were particularly strong in the historical urban core of the city, specifically in the districts that were devastated by the 1985 earthquake. This existing infrastructure of grassroots political mobilization quickly became a major factor in response to the earthquakes of September 1985 and helped to shape the entire plan for reconstruction. As the temblors significantly disrupted communication and other government functions, the public rapidly mobilized to fill in the void left by government incapacity (Poniatowska 1995).

In the earthquake, losses were spread across social classes due to the severe impacts on modern high rise structures that housed or were employment sites for middle-class and working-class people. There were also significant losses to the urban poor located in older, dilapidated tenement buildings. The vulnerability of the urban poor was closely linked to their very limited income opportunities that tied them to specific neighborhoods. As the earthquake destroyed many of the small businesses that provided marginal incomes, a precarious economic situation before the earthquake became critical for many households afterward (Kreimer and Echeverria 1991). Because the earthquake displaced more than 50,000 and destroyed hundreds of small business, restoring housing and workplaces became essential to avoid a deepening crisis.

In urban earthquakes in poor countries, the housing issue is often very difficult to resolve due to lack of national resources, political factors, and the fact that the housing problems are reflective of profound economic troubles that impoverish significant portions of a population. Following the Mexico City earthquakes, there were large-scale protests where citizens demanded government responsiveness to the needs of the victims. In response to this multi-class movement, the government ultimately agreed to rebuild new housing in place and to situate temporary housing on available sites in each neighborhood. Rather than be relocated to the urban periphery, households would be able to remain in neighborhoods that they had significant social and economic ties to as reconstruction proceeded.

The government, bowing to the grassroots insurgency, developed a reasonably democratic Popular Housing Reconstruction Program (HRP) (Stolarski 1991). Utilizing international funding, the program expropriated privately owned land on which the many heavily damaged apartment structures were built. The HRP worked with neighborhood associations, political movements, and NGOs, developing plans to rebuild damaged housing on site in culturally appropriate and aesthetic forms and to sell the new apartment units both to individual purchasers and condominium associations formed in the neighborhoods. This was accomplished in a way that was responsive to popular inputs, seeking cooperative relations with the earthquake victims who would be deeded the new residences on completion (Kreimer and Echeverria 1991). In all, 46,000 new or rehabilitated housing and commercial units were made available within two years of the earthquake, allowing victims to acquire more secure residences in neighborhoods many had occupied for more than a generation (Stolarski 1991). The Mexico City example illustrates the value of adequate funding applied creatively and of social democratic and participatory approaches to urban reconstruction after earthquakes. It also provides evidence of how historical conditions and political factors in civil society can shape the post-disaster situation. While the poor are often vulnerable to the impacts of earthquakes, mobilization of the marginalized and disenfranchised people in the wake of disasters can become a critical recovery resource (e.g. Anderson and Woodrow 1989; Poniatowska 1995).

Substandard housing and poverty may be associated with Third World earthquakes but their effects are frequently visible in US earthquakes as well. Several studies of recent California earthquakes have raised issues about political and cultural marginalization that produce inequalities in impacts, access to relief, and household recovery (Bolin 1994a; Phillips 1993). A number of US studies have identified social class, age, and ethnicity as important factors in recovery after earthquakes. An early study focusing on social inequalities in disaster recovery (Bolin and Bolton 1986) found low income Mexican-Americans to have experienced disproportionately high losses to their homes and possessions after a 1983 earthquake in Coalinga, California. More importantly, they had more difficulties obtaining relief assistance, lacked insurance protection, and had few financial assets to draw on to facilitate reconstruction in comparison to Anglos. To cope in the absence of adequate assistance from federal or state governments, Mexican-Americans relied on kin networks to augment their financial resources and to assist in reconstruction (Bolin and Bolton 1986).

Similar effects of both class and ethnicity on earthquake vulnerability were documented following the 1989 Loma Prieta earthquake in Santa Cruz County and the San Francisco Bay area. The Loma Prieta research (e.g. Bolin and Stanford 1991; Comerio et al. 1996; Phillips 1993) discussed the effects of a housing shortage on low income victims. In addition to 5,000 single-family homes, the earthquake heavily damaged or destroyed more than 4,700 lower cost apartment units in San Francisco, 1,300 in Oakland, and 1,000 in Santa Cruz, creating a housing crisis for the poor and marginally housed (Renteria 1990). Federal assistance programs failed to address the needs of those with chronic housing problems, particularly low income elders, the homeless, farmworkers, and poor African-Americans. Locally based programs, which attempted to address housing problems for vulnerable groups emerged at several locations in the impacted areas. The city of Watsonville saw significant political mobilization by displaced Latino farmworkers as they challenged the local Anglo power structure in the aftermath of the earthquake. The mobilization empowered local groups, rearranged class and ethnically based political coalitions, and led to new housing assistance programs (Laird 1991; Schulte 1991). In a fashion similar to political mobilizations of the dispossessed in Latin American disasters, the Watsonville movement is an important example of grassroots organizations working in conjunction with community-based organizations and disaster NGOs (Bolin 1994b; Phillips 1993).

In a study of low income Latino neighborhoods after the 1987 Whittier-Narrows earthquake (Los Angeles County), Bolton and her colleagues (1992) documented the difficulties of the urban poor in accessing federal disaster assistance and in finding replacement housing. As with other California disasters, the situation was exacerbated by the pre-existing shortage of low income housing. Because urban tenement buildings are typically owned by absentee landlords, often in ownership consortia, they are slow to rebuild (if they are rebuilt at all). And because of seismic retrofit programs mandated in Los Angeles, substandard buildings must be rebuilt to current codes, resulting in higher housing costs that may exclude poor households. Various housing relocation services were developed by government agencies and NGOs to assist low income households in the search for new affordable housing in the East Los Angeles area. As most of the displaced had strong social connections to their neighborhoods, they were reluctant to relocate elsewhere in the search for affordable housing. The Whittier-Narrows earthquake provided significant evidence for the ethnic and class-based nature of social vulnerability in the Los Angeles metropolitan area after a moderate earthquake (Bolin 1994a; Rubin and Palm 1987). While none of these studies place vulnerability in historic context, their focus on structured inequalities links them to behavioral response studies popular in the US and vulnerability analysis as we have described it here.

The Kobe, Japan earthquake of 1995 provides evidence of the role of vulnerability in First World contexts. While Japan has advanced seismic standards for earthquake resistant construction, neither those mitigation measures nor its level of social preparedness in the Kobe region were adequate to prevent the heavy human toll in the disaster. With more than 6,400 dead, 35,000 injured and 300,000 homeless as a result of the earthquake and fires, Kobe demonstrated that wealthy, technologically advanced countries of the North are not immune to catastrophic disasters (Tierney and Goltz 1996). As in the case of Mexico City, the Japanese government failed to adequately assess or respond to the emergency in the initial days of the disaster. In the case of Japan it is all the more surprising given the high levels of organizational and technical expertise from its long history of earthquakes, typhoons, and tsunamis. The earthquake produced widespread disruption of water lines, natural gas, sewage, and electric lines, causing the port of Kobe to shut down for a period of months. Deaths occurred disproportionately among women and elders who were at home at the time of the main shock. In virtually all age groups, more women than men died in the disaster (IFRCRCS 1996). While there was a large-scale mobilization of volunteer groups to assist, local and national government agencies lacked any systematic way to incorporate the 630,000 volunteers into the emergency response as Japan lacks an organized system of voluntary organizations for disaster assistance (Tierney and Goltz 1996).

Sheltering and providing support services to several hundred thousand earthquake victims required 1,200 emergency shelters established in a large variety of surviving structures in Kobe (Sugiman *et al.* 1996). In a densely settled urban area, available land and buildings for emergency shelter and temporary housing can be difficult to acquire, particularly when damage to buildings and transportation links is extensive. While the government of Japan is currently providing temporary shelters to refugees, the long-term reconstruction of housing is moving rapidly although still incomplete (as of 1997). Preliminary evidence suggests that low income persons, elders, and the disabled may not be able to afford to move back into their ancestral neighborhoods, becoming permanent residents of the 'temporary' housing camps (IFRCRCS 1996). An analysis of the emergency period in Kobe indicates that levels of social preparedness were inadequate to such a large-scale disaster in a complex urban setting. Social protection afforded by building and land use left some groups more exposed to physical harm than others. Lastly the government lacked adequate damage assessment capabilities, emergency management organization, and recovery planning for a catastrophe of this scale in the region where it occurred (Tierney and Goltz 1996). Thus a very wealthy country with advanced mitigation measures nevertheless experienced large-scale material losses, while its social protection measures failed to prevent some groups from suffering disproportionately in the aftermath. An over-reliance on technical solutions and physical engineering led to a neglect of the social aspects of preparedness in the Kobe disaster (IFRCRCS 1996).

SUMMING UP

Although we have covered considerable ground in this chapter, we recognize that there is much in the literature that has not been discussed. Our interest has been on highlighting some of the dominant themes and theoretical directions in disaster studies. We have characterized the US research tradition as including older collective behavior and functionalist strands and as being overly concerned with whether or not disasters should be considered as social problems (in a sociological sense). If there is a unifying theme in traditional US approaches, it is the search for abstract universals regarding social action in disasters. The search for 'universals' is too often divorced from history and geography. This is partly a consequence of the use of systems language which leads to a neglect of human agency and human experience in specific socio-political contexts.

Vulnerability analysis directs attention to the root causes of people's vulnerability in specific contexts as factors that prefigure disaster. It concerns the political economic forces that structure people's lives and the various household coping strategies that are activated during crises. It situates disasters in the larger frame of history and in the conditions of everyday life, and it asks that we look beyond the physical 'event' as the cause of disaster. Questions of differential vulnerability and its causes are important in both Third and First World disasters, although the approach has been developed to analyze Third World calamities. However the US examples reviewed demonstrate that questions of vulnerability pertain even in settings with advanced measures of social protection and high per capita incomes. In the following chapter, we consider the historical factors that have helped shape vulnerability and security in California.

NOTES

1 The exception may be found in the November 1995 issue of the US journal *Mass Emergencies and Disasters*. That issue featured efforts by an international group of disaster researchers to define what a disaster is along with a commentary and critique by Hewitt on those statements. US anthropologists do cite vulnerability literature as illustrated in a recent review of research on disaster (Oliver-Smith 1996).

2 It is traditional, at least among sociologists in North America, to identify Samuel Prince's study of a munitions ship explosion in Nova Scotia, Canada, in 1917 as the first social scientific study of a disaster (e.g. Scanlon 1994a).

3 We include in the general category of 'system response' the considerable literature on complex organizations and their responses to disasters. The DRC has produced an extensive body of research on organizations during emergencies, much of which has been systematically analyzed by Kreps in a series of publications (e.g. Kreps and Bosworth 1994). Recent developments in the field seem to be taking it in a more holistic direction with an increasing focus on sustainability (e.g. Mileti 1997), thus our characterization is intended to describe the established research base.

4 In practice, collective behavior theory provides no specific independent criteria by which to distinguish 'collective behaviors' from other types of social actions, beyond the imposition of the category by the observer. Critics contend that the behavior theory places a reductionist emphasis on psychological states of individuals and assumes that participants in collective actions often act irrationally (Gusfield 1994; McAdam 1997). Disaster sociologists who identify with collective behavior theory reject this and emphasize the adaptive and rational aspects of human behaviors in disasters (e.g. Aguirre 1994).

5 The statement is a close paraphrase of Fritz (1961: 655) in a three-decade-old functionalist social problems text. As has been noted by numerous critics, this definition ignores the ways existing social processes in a given society are implicated in the forms the disaster takes (e.g. Quarantelli 1987). Interestingly, the United Nations Disaster Relief Organization uses a closely paraphrased version of Fritz in its definition of disaster (see Smith 1996: 20).

6 There is a vast literature on famines and an equally large number of theories to explain the causes and consequences of specific historical famines. Reviews of this literature include Shipton (1990) and Downs *et al.* (1991). Famine is also a frequent topic of the British journal *Disasters* and it is an excellent source of the most recent research (e.g. Hendrie 1997).

7 To avoid any confusion we note that in the US 'entitlement' is often used to refer to social welfare programs that are offered to certain fractions of the population by the federal government. Entitlements in the US usually encompass employment-based security programs that pay retirement pensions or unemployment benefits to those who have worked and paid in to a fund. Public assistance (welfare) programs, on the other hand are means (i.e. income) tested and only those who qualify due to specific conditions of deprivation may receive relief. The relief is viewed as charity paid for by others. Welfare programs are supervised and recipients generally give up their right to privacy in order to be monitored for continuing qualification. In this sense US disaster relief programs are more akin to welfare programs than traditional entitlements in that most are means tested and applicants are subject to investigation and monitoring. Entitlements is sometimes used pejoratively to refer to all social welfare programs. Sen's use of the term is considerably broader than the restrictive US use, encompassing a general set of rights and obligations that people have through economic, social, and cultural practices. In this text we use entitlements in Sen's sense, not the US sense. In the US entitlement and public assistance programs are being subjected to concerted political attack (e.g. Demko and Jackson 1995). Prime targets have been programs that provide food to the needy, health assistance to low income elders, and living assistance to female-headed households with young children (Abramowitz 1996).

8 The siting of hazardous waste facilities near low income ethnic communities is a form of environmental racism (Bullard 1990 and 1993). We recognize that this occurs on an international level in an intensified form. Increasingly multinational corporations locate hazardous production facilities and waste sites in Third World countries to take advantage of laxer environmental regulations, low wages, and limited workplace safety regulations (Johnston 1994; Merchant 1992). The disastrous leak of toxic chemicals that killed thousands in Bhopal India in 1984 is simply the most dramatic example of a widespread process of subjecting poor people of the South to work and environmental conditions that would be illegal in the First World.

SITUATING THE NORTHRIDGE EARTHQUAKE

A vulnerability approach to disaster seeks the roots of disaster in the political and economic factors that shape material life and landscapes of communities. Our discussion of the Northridge disaster begins with the larger contexts of California and the Los Angeles region, examining the historical-geographical factors that have produced its contemporary landscapes. This chapter provides context at three levels; first, by briefly examining the history of California's urban development, next, by tracing the evolution of policies about earthquake hazards and their effects on social protection, and last in an overview of events associated specifically with the Northridge earthquake. As we move among these different levels of understanding, the interplay of private interests and public policies emerges as a theme in risk and vulnerability as social products.

We begin by discussing the unlikely urban development of the seismically active deserts and arid foothills of Southern California into the geographically most expansive metropolitan area in the United States. The urbanization process is the result of extensive land speculation and real estate development driven by private capital but facilitated by large public expenditures to provide water. We next review how the earthquake risks entailed by this space-consuming developmental trajectory became the subject of federal and state hazard mitigation policies that are today very much part of the political organization of hazard management in California. Last, we begin to examine the general features of the Northridge disaster, reviewing both the patterns of losses and the organized responses to it. We develop our understanding of the disaster by conceptualizing it as a product of the history of the region, the nature of social protection in place, and the physical particularities of its seismic hazards.

HISTORICAL CONTOURS OF DISASTERS AND DEVELOPMENT IN CALIFORNIA

California's earthquake risk was produced by Anglo-European settlement and subsequent large-scale urbanization. Rapid population growth in its two primary urban centers, San Francisco and Los Angeles, has situated more than half the state's thirty million residents in areas of high seismic activity (Bolt 1993). Population growth and urbanization in both metropolitan areas, in turn, have been dependent on the acquisition and development of water supplies for domestic and commercial uses. In this section we trace the development of Los Angeles, from a tiny pueblo of a few hundred people in 1781 to a late twentieth-century postmodern megalopolis.

Water, power, and politics

While San Francisco's population boomed with the discovery of gold in 1849 in the Sierra Nevada a few hundred kilometers to its east, Los Angeles was to remain, for much of the nineteenth century, an agrarian backwater, a mix of old Spanish land grant ranchers and Mormon farmers (Reisner 1993). Urban growth in the arid south of California was historically constrained by the meager and intermittent water supply available in the Los Angeles river. As Kahrl (1982: 1) observes, 'The history of California in the twentieth century is the story of a state inventing itself with water.' Water became a valuable commodity, captured and distributed in publicly funded developments and used to promote expansive urban growth in Los Angeles and San Francisco. As with much else in the political economy of California, the commoditization of water would prefigure patterns of development in the remainder of the dry western US states. To provide adequate water supplies required public funding of major water works, a political process that has been a persistent feature of the last eighty years of California's history (Gottlieb and FitzSimmons 1991).

San Francisco, in wetter Northern California, was the vanguard of nineteenth-century urban growth in the western half of the US. Initial urbanization in the San Francisco Bay Area was rapid, initiated by the mid-century discovery of gold and a mass influx of people to work the gold fields. 'Out of practically nowhere, a formidable capitalist presence emerged along the Pacific Ocean rim of the New World, beginning a Californian tilt to the global space economy of capitalism that would continue for the next century and a half' (Soja 1989: 190). Facilitated by its natural harbor, the city rapidly grew into the west coast center of mercantile capitalism by the 1860s. Between 1848 and 1851, San Francisco's population grew forty-fold, to 35,000. In two more years it surpassed 50,000, moving it among the top twenty US cities. In the next fifteen years it grew into a financial center to rival New York City, fueled by commercial fishing, international trade, agricultural production in the Sacramento Valley, and a transcontinental railroad system linking it to eastern markets (Reisner 1993). By 1870 San Francisco had become one of the most vibrant capitalist centers in the west and California's primary nineteenth-century industrial and commercial hub (Soja 1989). Due to its rapid growth San Francisco was outstripping its limited water supplies and delivery system, raising the specter of growth limits in the twentieth century. By the early 1900s San Francisco's population was nearing 400,000 and the city boasted several steel frame high rise buildings scattered among its more common unreinforced brick buildings and wood framed structures (Bolt 1993).

By 1900, the resource strategy being pursued by the political and economic elite of San Francisco was the acquisition water rights in the public domain, via concessions by the federal government. The cornerstone of San Francisco's plan was to build an aqueduct from the Sierra Nevada foothills (Figure 3.1), to transport water captured behind a proposed dam in the Hetch Hetchy Valley, part of the pristine wilderness of Yosemite National Park. The effort to dam the Tuolumne River that flowed through the valley engendered immediate opposition from the Sierra Club, and a decade-long political battle was joined to stop the proposed dam and aqueduct. The battle over the dam prefigured decades of conflict over federal dam building programs in the western US (Gottlieb 1993). In the midst of the Hetch Hetchy dispute, another environmental factor emerged that would give more weight to San Francisco's claims of need for federal water.

On 18 April 1906, the San Andreas fault near San Francisco suddenly ruptured and the west side of the plate boundary jerked several meters northward (Bolt 1993). In the minutes that followed, the 'pseudo-Parisian splendor' of San Francisco (Reisner 1993: 60) had crumbled. As water-mains ruptured, fires burned out of control for three days. Fire fighters lacked pumps necessary to draw water out of the Bay so they resorted to dynamiting standing buildings to deny fires fuel and control their spread (Harris 1990). In the end, much of the city lay in ruins; unreinforced masonry buildings collapsed from ground motion and liquefaction, and wooden buildings were reduced to ashes in the fires that swept the city. The legacy of the earthquake was used politically to overcome opposition to the Hetch Hetchy dam and the city moved forcefully to acquire Tuolumne River water in the name of safety.[1] And while the disaster was used to leverage water resources from the federal government, the earthquake's effects were downplayed publicly so as to not discourage investment in the Bay Area by eastern capital (Mileti and Fitzpatrick 1993). Thus, the placement of several million people and an elaborate urban infrastructure on the heavily faulted topography of the San Francisco Bay Area was made possible by the publicly funded 250-kilometer aqueduct that began providing water in 1923 (Nelson 1983).

Los Angeles evidenced a very different growth pattern from San Francisco's, although it too appropriated water from distant sources. Its greater aridity, lack of a sea port, geographic isolation, and minimal productive base made it, in 1850, an improbable site upon which to build an empire of capital. The initial Spanish establishment of the Pueblo de Los Angeles in 1781 situated the village at a point where the Los Angeles river flowed at the surface year round. The Pueblo de Los Angeles was seized from Mexico in 1847 by US forces and the entire state was ceded to the US in 1848 (Zinn 1990). With the military conquest of Mexico complete, the stage was set for Southern California's water problems to begin as Mexican agricultural and ranching activities were soon supplanted by other developments. During the first decade of US control the water supply system of Los Angeles was adequate given its population (1,600) and agricultural economy. However, as Los Angeles slowly expanded, its old Spanish open-ditch water system became a fetter on future growth. And nowhere in the Los Angeles basin was there the water necessary to sustainably support a metropolitan region of more than ten million persons that would develop in the coming century. Until adequate water resources for an imagined future could be developed, Southern California's earthquake risk would remain minimal.

The Los Angeles basin had one asset that San Francisco did not: large amounts of easily accessible land. It was Los Angeles' ranching and agricultural potential that attracted the interest of San Francisco capitalists in the 1860s, particularly the fact that citrus crops could be grown in its warm Mediterranean climate (McWilliams 1976). Once Los Angeles was connected to the rest of California and the US by railroads in 1867, a barrier to its growth was removed. However, periodic droughts were ongoing reminders of the problem of water in the south. The drought of the 1860s bankrupted many cattle ranchers, and large areas of former Spanish land grant ranchos were bought up by San Francisco capitalists intent on land speculation. The age of real estate speculation was beginning in Los Angeles, an age that has continued unabated from 1870 to the present (Davis 1995; Soja 1996b).

In the late 1800s Los Angeles and its adjacent settlements relied on a combination of pumping ground water and capturing surface flows in reservoirs (Nelson 1983). As the mountains surrounding Los Angeles were deforested to provide wood for construction, what were once perennially flowing streams increasingly became subject to winter floods, while ceasing to flow all together in the hot dry summers (FitzSimmons and Gottlieb 1996). While the San Francisco Bay Area could look to the large snow-fed rivers flowing west out of the Sierra Nevada for water, Los Angeles had little to choose from locally. A population boom in the late 1880s was prompted by low train fares, an extensive advertising campaign by railroad companies for Los Angeles, and offers of cheap and idyllic home plots. The corporations owning the railroads were also among the largest private landowners in California and were intent on promoting growth. Thousands of migrants from eastern and midwestern states were attracted to this virtually unknown corner of the US. As Davis' account (1992) shows, in 1887 alone $100 million in land sales occurred in the Los Angeles area ($2 billion in current dollars), and sixty new towns around Los Angeles were platted for development. The boom was based on frequently fraudulent practices; plots of land being sold sight-unseen to eastern investors were 'in the bed of the Los Angeles river, or up the nine-thousand-foot summits of the San Gabriel Range' (Reisner 1993: 55). The boom was quickly followed by a collapse as Los Angeles' population dropped by half to 50,000 by 1892. And then oil was discovered in the scrub-covered hills of Hollywood and a new boom was on by the turn of the century (Soja 1989).

By 1900 a powerful ruling elite had taken form in Los Angeles and began to lay the groundwork to make it the premier urban center of the west coast (Dear 1996). Such ambitious development demanded the construction of an all-weather sea port, which Los Angeles lacked, and it required new supplies of water. Both the development of harbor facilities at San Pedro and Long Beach, and the acquisition of water would be accomplished with significant financial assistance from the federal government. Providing water revolved around complex and seemingly Byzantine schemes, endless legal challenges, and monumental public works programs (Gottlieb 1988). Los Angeles, through an aggressive municipal annexation program rapidly eclipsed San Francisco in the twentieth century to become the largest city in California.[2]

To provide water for Los Angeles, William Mulholland, head of the Los Angeles City Department of Water proposed, in 1905, the diversion of water via an aqueduct out of the Owens River Valley (Figure 3.1). This remote valley was situated in the high deserts (1,500 meters) on the eastern escarpment of the Sierra Nevada some 400 kilometers from Los Angeles (Nelson 1983). In 1907 public funds for the project were approved by voters and the construction of the Los Angeles aqueduct began. In 1913, ten years before the completion of San Francisco's aqueduct from the Tuolumne River, the Los Angeles aqueduct was completed and the diverted Owens River water began flowing to impoundment reservoirs in the still agricultural redoubt of the San Fernando Valley. The desiccation of the once fertile Owens Valley had begun, as surface flows began to be diverted south to Los Angeles and its expanding suburbs (Gottlieb 1993).

The Owens River water was directed south in increasing amounts and a number of reservoirs were built in the mountains of Los Angeles County to store it for future use, including one in San Francisquito Canyon near present-day Santa Clarita. The Saint Francis dam held back a reservoir above the channel of the Santa Clara River, the major drainage into Ventura County (Figure 3.2). Shortly after Mulholland inspected leaks and declared the new dam safe, it collapsed, triggering one of the most deadly US floods of the twentieth century (Outland 1977). More than 450 people died in the catastrophic flood that followed and Mulholland's career was abruptly terminated (Rogers 1995). The flood cost the city of Los Angeles more than $15 million in reparations to survivors of the disaster (Hundley 1992). This disaster notwithstanding canyon reservoirs still forms a key part of the metropolitan area's water storage and distribution system today.

By 1928 Los Angeles was actively pursuing new water to supplement its Owens Valley sources. Los Angeles' seemingly relentless search for water was never based on current usage but rather on anticipated future growth

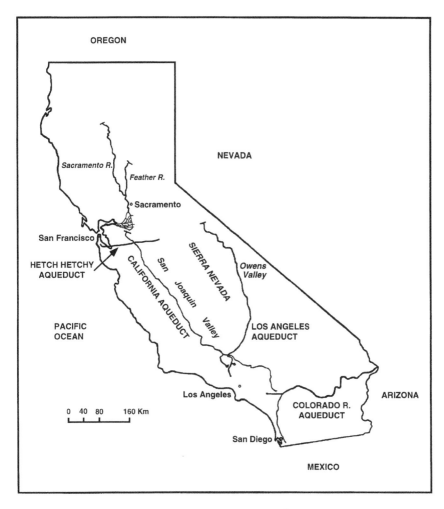

Figure 3.1 Major aqueducts and river systems in California

(Gottlieb and FitzSimmons 1991). The 1930s would inaugurate a forty-year era in large-scale federal water development projects in the western states. The first major project, one that would directly benefit Los Angeles, was the building of a 200-meter-high concrete dam on Colorado River near Las Vegas, Nevada. Its purpose was to generate electricity and provide irrigation water for the arid southwestern states. Along with the dam, the federal government developed a water allocation scheme, distributing impounded water to the seven states that bordered the river, with the largest share going to California (Gottlieb 1988). That allocation is pumped through Southern California's second major water lifeline, the 390-kilometer Colorado River aqueduct.

69

While the Los Angeles aqueduct is run by the LA Department of Water and Power, Colorado River aqueduct water is managed by Metropolitan Water District (MWD) which sells water to participating water agencies in Southern California (FitzSimmons and Gottlieb 1996). In 1960, yet another major publicly funded water diversion project was initiated. The State Water Project through dam building and canal systems began sending water from Northern California south to supplement MWD supplies (Hundley 1992). This intrastate diversion was facilitated by the construction of the California aqueduct which began delivering northern California water to the suburbs of Southern California in 1971 (Gottlieb and FitzSimmons 1991).

Facilitated by its successful water imperialism, by 1925 Los Angeles had a population of more than one million and it was growing at a faster rate than any other urban center in the US. By the start of World War II the Los Angeles region had 3.25 million in residence, and in the next fifty years it would add seven million more (Dear 1996; Davis 1992). In many ways, the sheer size of the population is less important than the urban form of the metropolitan area (see Figure 3.2). The fragmented and spatially dispersed style of urbanization emblematic of Southern California is of interest both in terms of exposure to hazards and in its environmental impacts.

Urbanization and hazardous environments

While the Los Angeles area has not had a major (greater than magnitude 7.0) earthquake in this century, it has a historical record of numerous earthquakes of sufficient magnitude to damage buildings and disrupt lives. In 1855, during the early period of 'Anglo' Los Angeles, an earthquake is reported to have damaged every building in the pueblo (Nelson 1983). Two years later, as a result of the M8.0 Fort Tejon rupture of the San Andreas fault 100 kilometers to the north, the Los Angeles area sustained further damage as a result of strong ground motion. Yet with a few thousand people spread out in small self-sufficient villages, one would be hard pressed to refer to that massive earthquake as a natural disaster. In 1872 a major earthquake (M8.3) in the Owens Valley was felt strongly in Los Angeles. The small town of Lone Pine, near the epicenter had some fifty homes destroyed and Bolt (1993) lists the death toll at about fifty. If there was an awareness of earthquake hazards in Los Angeles, it appeared to have no appreciable influence on the urban growth or planning in the region.

The only significant earthquake Los Angeles experienced prior to 1971 was a M6.4 event centered under the city of Long Beach, near the Port of Los Angeles, in 1933. Its source was the Newport-Inglewood fault, which runs under much of western Los Angeles County near the urban core of the city (see Figure 1.1). In spite of the moderate magnitude, damage was extensive, particularly to the many unreinforced masonry (URM) buildings clustered in this southern area of the county. Although the San Francisco earthquake

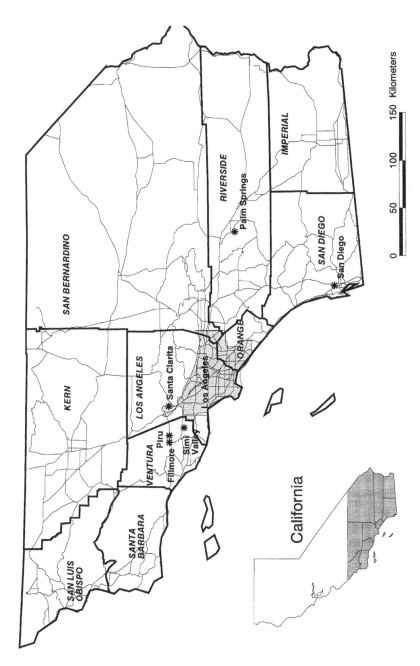

Figure 3.2 Major political jurisdictions in Southern California

had shown the seismic weaknesses of URM buildings, that knowledge was not reflected in local building codes of the period. Damage was significant in the city of Long Beach as well as in nearby municipalities. Compton, in particular, suffered extensive damage to its business district as a result of the liquefaction of alluvial soils. Approximately 120 persons died in the earthquake and property losses were estimated at $40 million (Harris 1990). The collapse of many school buildings led the California State Legislature to pass the Field Act one month later. The act mandated the seismic reinforcement of damaged school buildings and the incorporation of seismic codes in new school buildings and was the first of many state ordinances to enhance building safety (Mileti and Fitzpatrick 1993). The city of Los Angeles also enacted general building code changes for structures built after 1933 to reduce future earthquake related building collapses. Little else was systematically done in California (or nationally) regarding seismic safety policy for another forty years (Palm 1995).

At the time of the 1933 earthquake, the three million plus inhabitants of the Los Angeles area were dispersed over a network of suburbs, linked to distant water sources by aqueducts, and dependent on an increasingly elaborate network of highways and railroads to move about. Los Angeles' pattern of areally expansive growth anticipates urban development and the growth of suburbs in the rest of the US by decades (Garreau 1991). In analyzing this process, Harvey writes:

> Long reduced to a commodity, a pure form of fictitious capital, land speculation had also been a potent force making for urban sprawl and rapid transitions in spatial organization . . . [i]t took the rising economic power of individuals to appropriate space for their own exclusive purposes through debt-financed homeownership and debt-financed access to transport services (auto purchases as well as high-ways), to create the 'suburban solution' to the underconsumption problem.
>
> (Harvey 1989: 39)

The term 'urban sprawl' was coined in the 1950s to refer to the spatial development of Southern California (Whyte 1958). With a ratio of 1.6 people per automobile (Davis 1992: 119) in 1925, the Los Angeles region was becoming automobile dependent in ways that the rest of the US would not approach until the 1950s. Automobiles, of course, were one 'solution' to moving about a spatially fragmented urban area (Bottles 1987). This process of spatial dispersal, land development, and privately owned automobiles 'meant the mobilization of effective demand through the total restructuring of space so as to make the consumption of the products of the auto, oil, rubber, and construction industries a necessity rather than a luxury' (Harvey 1989: 40). And if the use of private automobiles requires both consumption and pollution, by 1915 Los Angeles also had an elaborate urban electric rail system connecting its many far-flung suburbs. Unlike San Francisco, Los Angeles never municipalized the rail system, and in private hands it failed to keep pace with urban growth in the financially shaky 1930s. In 1945 that urban rail system was taken over by a General Motors and Standard Oil trust, and allowed to go out of business, leaving automobiles as the 'mass transit' alternative for the 1950s and beyond (Bottles 1987).

Urbanization in Southern California, from early in this century, saw the spread of people and structures over an area criss-crossed by numerous faults and flood-prone water courses connected by increasingly complex and inefficient grids of electricity, water, gas, and highways (Davis 1995). The history of the region has been punctuated by efforts to prevent this 'space covering' type of urbanization in the Los Angeles basin. Among the first to articulate a limited growth vision for Los Angeles were members of the Socialist Party of California who, from 1908 to its completion in 1913, actively opposed the construction of the Los Angeles aqueduct. Harriman, leader of the Socialist Party, 'offered an alternative vision of a democratic community organized to live and grow within its existing resource base, with real estate subdivisions organized according to a plan rather than through speculation' (Gottlieb 1993: 29). The socialists were narrowly defeated in the election for mayor of Los Angeles in 1911 and soon after were dispersed into the deserts north of the city. Other critics of Los Angeles' early twentieth-century spatial expansion could be found among urban planners (Davis 1992). Numerous reports by urban planners of the 1920s and 1930s questioned Los Angeles' growth patterns and the lack of public parks and beaches. By 1928, in the midst of its second oil boom, the metropolitan area had less than 1 per cent of its surface area devoted to public recreational uses, a geographical irony in a city that marketed itself as an outdoor recreation Mecca (Davis 1995: 4; FitzSimmons and Gottlieb 1996). To provide open spaces, planners attempted to use flood plain hazard zoning to provide open spaces along the Los Angeles River and its tributaries. However federal flood control dollars of the 1930s were not used to provide open space along flood plains of Los Angeles. Rather, always in pursuit of development, federal subsidies were used to channelize and pave the Los Angeles River, turning it into a concrete storm sewer and giving over its flood plain, not to parks, but to industrial development (Davis 1995).

Recent transformations

The spatial expansion of the Los Angeles metropolis continued with considerable vigor through much of this century, in spite of continuing local challenges to slow growth or change its sprawling character (Davis 1992). By the 1930s, the area was seeing a proliferation of suburbs rapidly displacing citrus groves and farm land. And while agricultural production continued to decline across the area, a Fordist heavy industry sector grew in its place (Soja 1989). Four major automobile manufacturers opened assembly plants in Los Angeles during the 1930s and a growing aircraft industry also located there. The region became a major center of steel and tire manufacturing at the onset of World War II, to support military actions in the Pacific theater. Los Angeles' persistent anti-union environment made it an obvious choice for industries seeking low cost labor during this period (Davis 1986). The industrialization of Los Angeles in the middle third of the century was a draw to African-Americans escaping apartheid in the southern US and in search of industrial jobs. Some 600,000 blacks moved to Los Angeles between 1942 and 1965, one of the largest internal migrations in US history (Soja 1996b). Anticipating the growth of a 1970s 'Sunbelt' style of post-Fordist industrialism in defense/aerospace and electronics, Los Angeles was already in the 1950s a center of high-technology production.

For our purposes it is sufficient to observe that the spatial distributions of the various industrial sectors have developed very unevenly, and as the old Fordist industrial sector declined throughout the second half of the century, so did the working-class neighborhoods that drew their livelihoods from the sector (Davis 1992). Each wave of industrialization and deindustrialization has shifted income opportunities, leaving virtual industrial wastelands of poverty and unemployment. The old working-class neighborhoods that were hardest hit included the predominantly African-American areas south of the downtown in what is known as South Central Los Angeles (Anderson 1996). Here deindustrialization of the working class was the local effect of a global capitalist restructuring in the post-1960s era and the move to off-shore production (Davis 1986; Harvey 1990).

The decline of the Southern California's industrial working class has been counterpointed by the growth of two other economic sectors: electronics and the apparel industry. By the mid-1980s Los Angeles would have twice as many workers in electronics and aerospace as the much celebrated 'Silicon Valley' south of San Francisco (Hayes 1989; Soja 1996b). Few of these jobs are accessible to the former industrial working class, much of the work is low wage and part time (Watts 1992) and most sites of production are located well away from Los Angeles' 'rust belt' of defunct industrial plants. The recent growth of the garment industry reflects the other 'low-tech' side of reindustrialization in Los Angeles. While the number of garment industry jobs grew dramatically in the 1980s, so did the number of undocumented immigrants employed in the industry. It is estimated (Soja 1989: 207) that as many as 80 per cent of the low-paying jobs are held by undocumented immigrants, and that 90 per cent of the jobs are held by women. In Soja's words (1989: 207–208): 'Sweatshops which provoke images of nineteenth-century London have thus become as much a part of the restructured landscape of Los Angeles as the abandoned factory site and the new printed circuit plant.' The economic landscape of Los Angeles is increasingly polarized between extremes of income, as well-paying industrial jobs have declined precipitously, replaced by low wage, part-time, and temporary work (Anderson 1996).

On the other hand, the downtown core of the City of Los Angeles has become in the last twenty years a new international center of banking and finance, with a new skyline of monumental architecture to accompany its status as a Pacific Rim 'capital of capital' (Soja 1996a). But if the urban core is a center of international capital and its high income managers, it is also a place with the largest homeless population in the United States, reflective of the growing housing crisis in Los Angeles (Wolch 1996). The elements of economic success and sectoral expansion are counterpointed by areas of economic decline and abandonment across the region. The restructured economy with its aggregate growth in jobs is heavily segmented according to gender, race, ethnicity, and immigrant status, producing a fragmented terrain of class and ethnic inequalities (Soja 1996b). To what extent such social inequalities and the 'at risk' populations they produce (Demko and Jackson 1995) may be implicated in hazard vulnerability will be examined in the following chapters.

The spatial and economic restructuring of the Los Angeles metropolitan area has been accompanied by significant demographic shifts in its population. In 1960 Los Angeles County was approximately 85 per cent white. Since this period of 'Anglo Los Angeles', the metropolitan area has become a polyethnic mix unrivaled in the United States. As Davis (1992) notes, Anglos became a numerical minority in Los Angeles in the mid-1980s and are projected to be a minority in the state sometime during the first decade of the next century. Yet for all its polyethnicity, Los Angeles also has had, until relatively recently, some of the most sharply segregated ghettoes and *barrios* in the United States (Mate 1982; Soja 1996a). Now, racial and ethnic segregation has begun a decline as a result of the transformations accompanying the high rates of immigration into the region. Ethnic desegregation has been produced from the diffusion of Latinos and Asians throughout the metropolis, adding to the ethnic mix of many municipalities (Rocco 1996).

The burgeoning multi-ethnic spaces of Los Angeles are shaped by continuous economic change that has shuffled and reshuffled production centers in the region. As many as two million immigrants have moved to Southern California since 1970, including a large migrant flow from Mexico. Soja (1989: 216–217) has characterized the new immigration as a 'reserve army of minority and migrant workers' that has created 'an overflowing pool of cheap, relatively docile labor that is not only locally competitive but also able to compete with the new industrial concentrations in the Third World'. This influx of often low income immigrant groups has put a severe strain on the availability of affordable housing in the Los Angeles region (Horton 1992; Wolch 1996). Those in the country illegally live under continuous threat of deportation if detected, and lack political representation or worker's rights.

A continuing feature in the outward suburban push in Los Angeles (and many other US cities) has been 'white flight' (Davis 1992), although it is as much about class as race. The pattern reflects the move outward away from older parts of the city as homeowners 'escape' from urban problems, real or imagined. Older neighborhoods are abandoned by those who can afford to move and the new migrants settle in new class-segregated housing developments ostensibly far from the urban troubles of their former communities. The San Fernando Valley is a prototype of the older suburban fringe, a white middle-class refuge of the 1950s built on citrus orchards and farm lands. But by the 1970s whites were no longer a majority in the San Fernando Valley, and the valley was taking on a decidedly polyethnic mix, bolstered by a large Latino population in the eastern section of the valley in the city of San Fernando (Rocco 1996). The latest phase of outward expansion has seen population movement north over the Santa Susana mountains into Santa Clarita and beyond in new 'edge cities' (Garreau 1991).

From the Santa Clarita Valley north across San Andreas fault rapidly growing suburban developments abound. Spillover growth is also creeping into the agricultural stronghold of Ventura County, in particular to Simi Valley, a few kilometers from the San Fernando Valley. Yet there are few major employers on the edges of Los Angeles and those who seek suburban safety on the margins face time-consuming commutes into the San Fernando Valley and points south for employment. In Davis' dystopic imagery of this expansion (1992: 12):

> Setting aside an apocalyptic awakening of the neighboring San Andreas Fault, it is all too easy to imagine Los Angeles reproducing itself endlessly across the desert with . . . pilfered water, cheap immigrant labor, Asian capital, and desperate home buyers willing to trade lifetimes on the freeways in exchange for . . . dream homes in the middle of Death Valley.

There are several features of this developmental process that bear on vulnerability and disasters. The first is the acquisition of water resources and the attendant influences that process has had on urbanization. Water, its control and delivery, are central in the evolving political terrain of California, and has directly influenced the concentration of urban populations in a hazard-prone area (Gottlieb and FitzSimmons 1991). Protecting an infrastructure that spreads out over thousands of square kilometers is a key issue in earthquake safety as is protecting the resident population that water makes possible (SSC 1995). The second is the political and economic processes shaping Los Angeles as a spatially dispersed, polynucleated urban region. This connects both to the hazardous environments that residents may occupy, and to the environmental degradation that accompanies development (FitzSimmons and Gottlieb 1996). With urban expansion and growing populations come concomitant engineering difficulties and public expense in protecting the region against damage from the apparently inevitable earthquakes. Another feature is the growing ethnic and socioeconomic diversity of the region that has accompanied successive waves of economic restructuring and population expansion. The sociocultural and class diversity increases the complexity of managing actual disasters and reducing the potential impacts of future events. This has to be considered in light of the growing income inequalities, rising housing costs, and decline of social entitlements that make some people's lives increasingly precarious.

With the infrastructure, lifelines, and lives of fifteen million spread out over the heavily faulted topography of Southern California, earthquakes are an issue of considerable concern at both the federal and state levels. The various state-led efforts at seismic safety have their own history, a history that connects first with Alaska and only later with California's own earthquake hazard.

EARTHQUAKES AND HAZARD MITIGATION POLICY

Federal interest in earthquakes came about only after the government first established a role in providing general disaster relief to communities. Federal disaster relief policies have evolved piecemeal over the last four decades, as the state has gradually expanded its authority in matters relating to hazards and disasters. Much of the federal policy making since 1950 has taken place in response to actual disasters, promulgated in congressional responses to particularly destructive disasters rather than to risks routinely posed by natural hazards (May 1985). A major disaster creates identifiable needs that political representatives can use to leverage federal assistance through specific legislation designed to assist the politicians' jurisdictions. At other times, hazards appear to recede in importance and little is done legislatively (Rossi *et al.* 1982).

The first significant piece of legislation for federal disaster relief was the 1950 Disaster Relief Act, which authorized assistance for the repair of local public facilities. Numerous additional federal policies were developed over the next fifteen years, leading to a revised Disaster Relief Act of 1966. With each amendment to federal policy, the government expanded the types of disasters it would assist in and broadened its areas of responsibility for assisting local governments. State and local governments understandably have been supportive of growing federal responsibility that provides them with financial assistance after calamities. Changes in policy, including the Disaster Relief Act of 1974 and its subsequent iterations, have been increasingly generous in assisting individuals, businesses, and local governments. The 1988 Stafford Disaster Relief and Emergency Act, the most recent significant legislation, established federal involvement in preparedness, mitigation, response, and recovery. Pursuing a redistributive policy (May 1985), the federal government has increasingly socialized the costs of major disasters in the US.

Political concern with earthquakes was scarcely visible in the decades following the San Francisco Earthquake of 1906. This disinterest connects to political and economic issues, not to scientific knowledge, or its lack, regarding earthquake risks. Federal seismic policies are the outcome of a political process in which the threat of earthquakes had to be framed as a *national* problem before federal actions could be initiated (Birkland 1996). It has been a contentious process with disagreements over the seriousness of the earthquake problem, how much should be done about it, who should do it, and who should pay for it.

Historical perspectives

California experienced numerous large earthquakes in the period prior to the enactment of federal earthquake legislation in 1977. Apart from the 1906 Great San Francisco earthquake, none of these produced a significant disaster and none was directly experienced by more than a small fraction of the state's population. As a result, most earthquake hazard reduction efforts in California have been in response to local experiences, as in the case of the 1933 Long Beach earthquake. More recently policy making has begun to respond to future risks rather than actual disasters. Claims about California's earthquake risk are based on what seismologists can decipher from the evidence at hand. At issue is their ability to convince others (citizens, politicians) that a problem exists that must be dealt with. Neither the Great San Francisco earthquake nor the Long Beach earthquake made it generally 'obvious' that California's urban centers were at significant risk from earthquakes. Rather, as Stallings (1995) shows, the earthquake risk has been socially constructed relatively recently by geoscientists and policy makers as new theories, particularly plate tectonics, and research methods provided a better scientific understanding of seismic risk. Much of that technical knowledge has been gained over the last thirty years as federal funding was made available for extensive earthquake research (Bolt 1993).

The critical event in generating federal attention was the M9.2 earthquake that damaged large areas of south central Alaska in March, 1964 (Mileti and Fitzpatrick 1993). Following the 1964 earthquake the federally supported National Academy of Sciences (NAS) began a study of the disaster. The resultant research, published in several volumes six years later (NAS 1970), included geological, engineering, and socioeconomic studies. It also marked the start of a quest in the US, still unfulfilled, for earthquake prediction capabilities (Mileti *et al.* 1981). In spite of increased federal attention to earthquakes, no federal policies were passed, and seismic hazard mitigation activities remained at state and local levels into the 1970s. Stallings (1995) suggests that one reason for the lack of federal action was that earthquake problems were seen primarily as an issue for states like California. To become a federal concern, the earthquake risk would have to be reframed as a national problem.

In 1971 a moderate earthquake in the San Fernando Valley, near the municipality of Sylmar, caused damages estimated at $1.8 billion (in 1997 dollars – see Figure 1.1). A number of highways and large buildings collapsed from the M6.4 temblor, including a federal Veteran's Administration hospital. A total of sixty-four deaths were attributed to the earthquake, approximately the same as the Northridge earthquake would claim more than twenty years later. According to Olson *et al.* (1988) the San Fernando Valley disaster became the key to the promotion of federal earthquake hazard policy. From 1971 to 1977 a US senator (Cranston) from California introduced a number of bills to establish a federal program for earthquake hazard reduction. During the same period the federal Office of Emergency Preparedness (forerunner to FEMA) began a push for a broad federal hazard reduction policy, to go with its existing disaster relief policies. Federal policy for flood control and mitigation, as noted, had already been in place for decades (May 1985) so there were ample precedents for federal hazards policies.

In spite of support from the US president (Nixon), who had recently enacted a comprehensive National Environmental Policy to address techno-logical hazards (Szasz 1994), no federal policy for earthquake hazard made it through the congress until 1977. Only when seismic safety advocates, including both politicians and geoscientists, convinced a congressional audience that the earthquake problem was indeed a national concern did federal involvement begin (Stallings 1995). The claims-making process was pushed along by deadly earthquakes in Guatemala and Tangshan, China in 1976. Adding to risk promotion, the USGS discovered and documented a 'bulge' near the San Andreas fault north of Los Angeles (Palmdale), which suggested that seismic pressures were mounting and an earthquake was imminent (Mileti and Fitzpatrick 1993). Evidence of earthquake hazards was assembled and presented in a national political arena (the US congress) and claims made that earthquake risks were widespread in the US. Enough members of congress were convinced and a bill (introduced by Cranston) was ultimately signed into law.

In 1977 the Earthquake Hazard Reduction Act became the first federal earthquake legislation in the US[3] (Olson *et al.* 1988). The National Earthquake Hazards Reduction Program (NEHRP) become the program for implement-ing the Act and the Federal Emergency Management Agency (created in 1979) was designated as the lead federal agency. NEHRP's primary mission is to enhance and promote earthquake preparedness and hazard mitigation (Mileti and Fitzpatrick 1993: 22). A key agency for NEHRP implementation is the USGS through its extensive research program to identify faults, to map seismically hazardous areas, and to develop forecasts or predictions about future earthquakes (Mileti and Sorensen 1990). Earthquake engineering is also supported in the program in efforts to improve the seismic resistance of buildings and other structures (bridges, lifelines, highways). The program was reauthorized in 1990 to update its mission statement, adding building code development and improving research applications. The technocratic approach predominates here: solutions to the earthquake risk lie in the hands of engineers, geoscientists, and policy experts; all that citizens can do is be prepared for earthquakes (Stallings 1995: 42).

Claims-making and earthquake policy

The federal government has no direct mechanism to force implementation of seismic safety provisions on states, although it does offer incentives, including funding opportunities for mitigation after disasters. The state of California and the City of Los Angeles lead the United States in adoption of seismically resistant building codes, hazard zoning for land uses, and public education for earthquake awareness (SSC 1995; May and Birkland 1994). A key actor in the state's 'earthquake community' is the Seismic Safety Commission, a governor-appointed commission that promotes earthquake risk reduction policy, both in the state legislative arena and with municipal governments.

Even in the case of California, with its historical memory of earthquakes, efforts to change land use practices and building codes do not necessarily go unopposed. In a state that has major economic interests in real estate development, housing construction, and commercial building, efforts by safety advocates to restrict development or to impose building codes that raise the cost of construction are often contentious (Olson 1985). As with the federal government, California has delegated authority for many aspects of hazard safety downward to county and municipal governments. Adoption by city and county governments of earthquake mitigation measures is a voluntary process, sometimes influenced by lobbying from earthquake safety advocates like the SSC (e.g. Alesch and Petak 1986). However, as the Seismic Safety Commission has observed, local governments do not necessarily promote or enforce seismic safety codes and rules (SSC 1995).

The California Office of Emergency Services (OES), the key emergency management agency in the state, is actively involved in implementation of earthquake safety plans and in public education activities. Its sponsorship of regional groups like the Southern California Earthquake Preparedness Project has been part of its earthquake safety implementation mission (Lambright 1985). Irrespective of the state and federal levels of support for earthquake safety, adoption of safety measures at local levels is often slow and not without political conflict. Local governments are often major promoters of real estate and commercial development in California (Davis 1992) and they do not necessarily want to implement policies that would curtail population and business growth. Political 'will' sometimes increases after an earthquake when local governments adopt new seismic measures or decide to enforce existing ones more assiduously (Birkland 1996). Such was the case after the 1933 Long Beach earthquake, which prompted several local governments to adopt more stringent building codes and led to state legislation to mitigate seismic hazards in schools, hospitals, and theaters (Mann 1979).

California approved a measure in the aftermath of the 1971 Sylmar disaster requiring hazard zoning to limit construction on active earthquake faults. The Alquist-Priolo Special Studies Zone Act was passed in 1972 mandating the development of fault zone maps by the State Geologist that identify areas subject to surface rupture during seismic activity. The Act requires that designated hazard zones be made public information and that real estate agents inform clients if their home purchase is located in a 'special studies zone' (Palm 1995). It was renamed the Alquist-Priolo Earthquake Fault Zone Act in 1993, doing away with its obscurantist euphemism (special studies) which failed to convey any sense of risk to home-buyers. Being located in an officially designated fault zone has no effect on property values or home prices (Palm and Hodgson 1992), suggesting the irrelevance of hazards in California housing markets. Recent earthquakes including Northridge and the Whittier Narrows occurred on blind thrust faults with no surface ruptures. In neither disaster were there any designated Fault Zone Act areas involved, pointing to an obvious policy limitation.

The City of Los Angeles has adopted the most rigorous seismic safety measures anywhere in the US and the most stringent in California. However, in spite of a supportive political culture, seismic safety proposals have been contested, resulting in protracted city council struggles to adopt and implement specific measures (Alesch and Petak 1986). Exemplary of such struggles is Section 88 of the City of Los Angeles building code mandating seismic strengthening of unreinforced masonry buildings built prior to the 1933 Long Beach earthquake. The policy process for Section 88 began two years after the 1971 Sylmar earthquake, when a Los Angeles city council member introduced an ordinance to require pre-1933 theater buildings in Los Angeles to either be seismically upgraded or demolished because of the hazards they presented (Alesch and Petak 1986). That ordinance drew immediate opposition from a variety of special interest groups and was defeated in 1975. A revised plan introduced in 1976 was likewise defeated by a coalition of chambers of commerce and the Apartment Association of Los Angeles County (a landlord organization). The ordinance went through a variety of permutations over the next few years to appease property owner objections. In 1981, seven years after it was first introduced, the ordinance was approved and became Division 88 of the Los Angeles Building Code, mandating phased retrofitting of URM structures (Olson 1985).

In 1986, the state of California adopted a similar measure, the URM Law, which requires all communities in the state to identify unreinforced masonry buildings and develop plans for hazard reduction. Many communities avoided mandatory programs in deference to building owners who could not afford the strengthening procedures. Fillmore, heavily damaged in Northridge, was a community that chose in 1993 not to make mitigation mandatory (SSC 1995). Other communities that made retrofit optional include Whittier, Santa Cruz, Watsonville, and Coalinga – all communities that have had major damage to URM buildings from recent earthquakes. Contention over Los Angeles' mandatory retrofit program is but one example of how hazard mitigation can be politicized if it threatens commercial interests or property rights.

Earthquake safety in California cannot be isolated from the political and economic forces that have shaped and reshaped its urban forms. It is an urbanism susceptible to having its dispersed infrastructure of highways, water systems, and power transmission grids disrupted in large earthquakes. The federal interest in California earthquakes has proceeded from the notion that a destructive earthquake which negatively impacts the California economy will ripple outward and disrupt the national economy (Palm et al. 1990). In Stallings' (1995: 187) view, managers who promote earthquake safety are acting as agents of the state in seeking to maintain the functioning of a troubled economy in the face of potential disruptions by disasters. In that sense, the basic issue is not earthquake safety per se, but the need to minimize disruptions to the overall accumulation process in the region. State management of complex urban areas is challenged under the disorganizing effects of actual disaster conditions (Hewitt 1997). Only the state can manage the built environment as a whole, yet it must balance between 'what is good for accumulation and what is good for people' (Harvey 1985b: 47). Earthquakes, and hazards in general, provide a political management arena in which doing things to protect civil society can also help armor the system of private production against untoward events. This is a double-edged sword in the sense that if the state does too little to reduce hazards, significant interruptions in accumulation may occur, and if it does too much, the costs of production may be raised significantly, and businesses relocate to less costly areas.

The space-consuming form of urbanism so prevalent today in many parts of the US is itself a consequence of federal and state government acquiescence to the interests of capital in promoting largely unregulated development (Harvey 1989). The situation in California could be characterized as one in which some government agencies try to prevent seismic disasters to a large, complex, difficult to manage urban form that other political interests have facilitated and promoted. Seismic safety as currently advocated is about the built environment and the protection of property and lives. The social conditions that place some people at risk, and the economic conditions that make some people vulnerable to a calamity are not addressed in any disaster reduction mandates. Northridge, more than any other recent earthquake, has been significant because it produced housing damage throughout the Los

Angeles region, it disrupted key infrastructural functions, and it demonstrated that 'state of the art' engineering standards may not be adequate in large earthquakes (SSC 1995). It is to these matters that we now turn.

A MODERATE EARTHQUAKE IN LA

The Northridge earthquake inscribed its effects across the politically and socially fragmented landscapes of Los Angeles, Ventura, and Orange Counties in 1994. It originated on an unknown blind fault 30 kilometers southwest of the 1971 Sylmar earthquake's epicenter. In the twenty-three years that separated the two temblors, population in the LA region had increased by two million and building codes and other seismic safety practices had been enhanced. Northridge was the first significant earthquake to strike the metropolitan region of Los Angeles since the 1987 M5.9 Whittier Narrows earthquake leveled Whittier's business district and damaged apartments across the *barrios* of East Los Angeles (Bolin 1994b). The Northridge earthquake revealed numerous weaknesses in the abilities of structures and lifelines to endure severe ground shaking (USGS 1996). It demonstrated strengths and weaknesses in emergency response capabilities of the various municipalities and counties (Nigg 1997). In the discussion that follows, we focus on the general physical effects of the disaster and the efforts by federal and state managers to restore lifelines and infrastructure. We note strategies and problems involved in managing a disaster in a highly dispersed and socially diverse metropolitan region.

An overview of losses

The Northridge main shock and the nine M5+ aftershocks produced zones of damage scattered across the Los Angeles region, extending into more rural areas of Ventura and northern Los Angeles counties (Figures 3.3 and 3.4). There were concentrated areas of significant damage in the epicentral region as well as other clusters of damage near Hollywood, Beverly Hills, Santa Monica, and in the predominantly Latino and African-American South Central[4] district of Los Angeles (OES 1995). To the north, the Santa Clarita Valley received extensive damage to its building stock as did the Ventura County communities of Fillmore, Piru, and Simi Valley (Figure 3.3). In total more than 56,000 residential units (mostly apartments) were heavily damaged or destroyed in the main shock and several powerful aftershocks. More than 19,000 single-family homes sustained damages in excess of $10,000 (Comerio 1995). The value of direct losses is currently estimated to be at least $25 billion although the figure has risen steadily for nearly two years (OES 1997). The earthquake prompted the largest federal expenditure for a disaster in US history, more than $12 billion in grants and loans by 1996

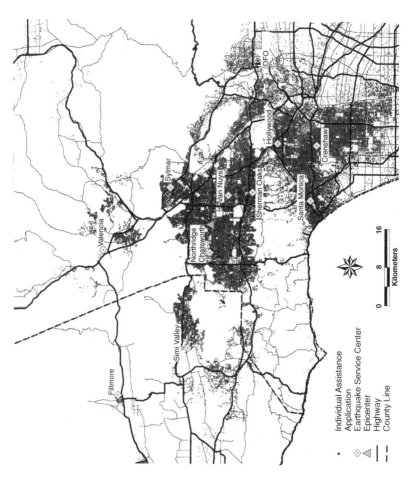

Figure 3.3 Locations of individual assistance applications in the disaster region

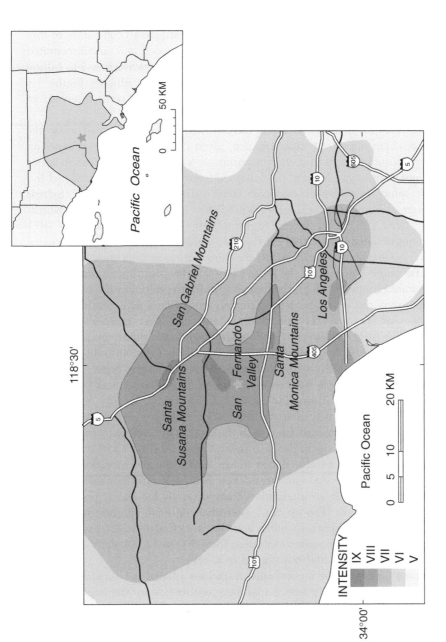

Figure 3.4 MMI zones in Los Angeles disaster region

(FEMA 1996). Insured losses for residential and commercial structures are currently at $12 billion making private insurance the single largest source of reconstruction funds (Comerio *et al.* 1996). The insurance pay-back produced a financial crisis in the insurance industry and prompted all companies to stop selling earthquake insurance in the state. Uninsured and uncompensated losses are now estimated to push total direct losses to as much as $44 billion (OES 1997), although a precise figure is elusive if not illusive. Figure 3.3 illustrates areas from which applications for assistance were received, an indicator of locations of damage.

Damages to lifelines and urban infrastructure were widespread across the metropolitan region. Of chief concern, given the overwhelming reliance on private automobiles for urban transportation, was the collapse of seven major sections of freeways. Most of the collapsed freeway bridges were designed and built before new seismic standards were instituted after the 1971 Sylmar disaster, although two bridges retrofitted after 1971 collapsed during Northridge (SSC 1995: 88). Several highway structures that survived the 1971 earthquake failed in the 1994 earthquake. The most serious collapses occurred in Gavin Canyon which cuts across the Santa Susana mountains separating the San Fernando Valley from the Santa Clarita Valley resulting in the closure of Interstate Highway 5 (I-5), the main north–south arterial in the state. Traffic access from the north became extremely constricted as a result. Elevated portions of Interstate Highway 10, a large east–west motorway through central Los Angeles, also collapsed causing significant traffic congestion. The collapsed freeway alone carries 330,000 vehicles per day, an indicator of the traffic burden shifted on to other roads with its closure. Some thirty-nine other highway bridges were severely damaged in Ventura and Los Angeles County, and an additional 200, although functional, needed repairs (SSC 1995). Only one fatality is attributed to the numerous highway collapses, a remarkably low figure associated with the absence of traffic early on a national holiday. Total losses to highway infrastructure exceeded $350 million, not counting the costs of traffic delays in lost labor time and increased fuel consumption from extended commutes. The disruptions in the freeway systems pointed to the lack of alternative transportation for the region's several million commuters. The existing interurban railroad, the Metrolink, with lines running into Ventura County and the Santa Clarita valley, experienced a short-term increase in usage, increasing from 900 to 22,000 daily commuters for a few weeks until highways began to re-open.

Other lifeline problems included numerous natural gas pipeline failures, although comparatively few actual fires resulted from the ruptures. One spectacular fire, accompanied by streams of water from broken water mains became a media icon of the disaster, figuring prominently in television and newspaper reportage. Breakages in gas lines produced fires that consumed more than 180 mobile homes, the single largest category of structures that burned (Goltz 1994). Electric service was disrupted after the earthquake for approximately two million persons, but was restored for the majority of users within twenty-four hours. The power failures were significant in that the earthquake occurred in the pre-dawn hours and the darkness added to the difficulties people faced in getting out of collapsed structures or rescuing those trapped within them. Unlike the rapid restoration of electric power, water service interruptions persisted in some areas for more than two weeks. Santa Clarita and Simi Valley had significant damage to their water purification and distribution systems. These peripheral areas also had to rely on emergency reserves as a result of a temporary shutdown of the Los Angeles aqueduct and four other major pipelines used by the region to import water (SSC 1995). Damage to water filtration plants resulted in parts of Ventura County having to issue 'boil water' orders to households for several weeks to protect against potential bacterial contamination. Damage to some of the major feeder lines for water took more than a month to repair. Another significant point of weakness in water supplies was damage to above-ground steel storage tanks, several of which collapsed, spilling their contents. One tank in Simi Valley collapsed flooding homes otherwise undamaged in the earthquake. Since these tanks represent immediate water supplies for local distribution systems, their destruction complicated provision of drinking water to the affected sites in Simi Valley.

Damage to commercial structures and residential units has been extensively documented in data bases from the California Office of Emergency Services (e.g. OES 1995). Building inspection data list approximately 114,000 residential and commercial structures were damaged in the three-county area. Los Angeles County reported more than 2,650 buildings destroyed (red tagged), and 10,500 seriously damaged (yellow tagged), containing 56,500 residential units.[5] An additional 87,400 Los Angeles County structures were inspected and 'green tagged'. (see Table 5.6) Ventura County, with one-tenth Los Angeles' population, had 342 structures red tagged, approximately 1,000 yellow tagged, and another 3,100 green tagged. (The incompleteness of OES data for Ventura County is reflected in the fact that their building damage data base lists 4,500 inspected structures in the county with 'unknown' levels of damage, approximately the same number as known damage levels – compared to 4,130 'unknowns' in Los Angeles County out of more than 100,000 inspected structures).

The greatest building damage in both counties was to residential structures, including apartment and condominium complexes, detached houses, and mobile homes. Seven times as many apartment units as individual houses were damaged in the earthquake. More than 49,000 apartment units were yellow or red tagged in contrast to 7,000 single-family dwellings in Los Angeles County, with the majority located in the City of Los Angeles in the San Fernando Valley. A total of 19,000 apartment units were vacated in the City of Los Angeles (Comerio 1995). Los Angeles County had 1,853 commercial structures damaged or destroyed to Ventura County's 124 (OES

1995). These commercial structures together housed thousands of individual businesses. Current estimates are that 23 per cent of the direct losses in Northridge are from earthquake-related disruptions of business operations (Tierney and Dahlhamer 1997).

In fifteen neighborhoods in the San Fernando Valley (City of Los Angeles), more than 60 per cent of the housing stock was lost, most in apartment complexes that had to be vacated. Some 15 per cent of the lost rental stock in the Valley was in areas with poor 'non-white renters' (Comerio 1995) although, in general, this suburban area lacks the concentration of poverty-level households found in other districts of the city. Areas that lost high per centages of rental housing came to be referred to as 'ghost towns', as most of the residents had to abandon the areas for replacement housing elsewhere (Stallings 1996). The loss of 19,000 residential units in the City of Los Angeles was softened somewhat by a recession-induced surplus of rental units in the San Fernando Valley. As discussed in Chapter 6, a problem in the designated ghost town neighborhoods was getting owners to repair or rebuild their apartment complexes, a matter which the City of Los Angeles pursued aggressively with special loan programs funded by the federal government (LAHD 1995). Comerio notes (1995: 43):

> In the months following the Northridge earthquake, the combination of pre-earthquake economic conditions, high rental vacancies, declining property values, declining rental income, high mortgage debt and severe reduction of owner equity made it clear that recovery would not take place without a significant public intervention.

One area of concern was with the performance of unreinforced masonry (URM) buildings in the main shock. Of the 5,900 retrofitted URM structures in Los Angeles County, 400 were damaged in the earthquake, and fifty of those were red tagged and razed. This suggests that the retrofit program has been largely successful in preventing building collapse and the attendant threats to life safety (USGS 1996). Most URM structures are located in older sections of Los Angeles south of the San Fernando Valley. Many of LA's URM structures are apartments housing very low income persons and any losses to housing stock of this type adds to the growing shortage of affordable housing in the region (Wolch 1996). Rental units built before 1978 have their rents con-trolled by the City of Los Angeles Rent Stabilization Ordinance. If lower income victims are forced out of their apartments from the earthquake, they run the risk of not being able to find other rent controlled housing (e.g. Bolton *et al.* 1992). In Ventura County, Fillmore experienced extensive damage to sixty-four URM structures, none of which had been seismically retrofitted (Tyler 1997a). Unlike Los Angeles, Ventura County had little surplus housing before the earthquake (a 1–4 per cent vacancy rate by municipality), and the loss of housing constricted the market further.

In addition to apartment complexes, mobile homes also proved susceptible to damage. More than 5,000 mobile homes in Ventura and Los Angeles counties were knocked off mounting jacks and damaged. The fact that many mobile home residents are lower income and pensioners raised concerns about safety standards in the mobile home parks. Unlike other types of housing which are regulated at the municipal level, mobile homes are regulated by the state Housing and Community Development department. The HCD took a lead role in the subsequent rehabilitation of the mobile home parks in LA and Ventura counties. Other significant areas of loss included thirty hospitals that were damaged in the temblor, with four of those requiring total evacuation of patient populations. Health care facilities are the subject of the 1972 Hospital Safety Act and, as with public schools, are required to meet enhanced building code standards in California. At least 106 school buildings in the region were damaged extensively and couldn't be occupied safely. The earthquake also triggered the collapse of multi-story parking garages throughout the region, including one at California State University, Northridge (SSC 1995). The collapse of these structures suggests that had the earthquake occurred when lots were in use, the death toll would have been considerable. The university parking garage was built in 1991 to conform with the most recent engineering standards.

In accounting for the extent of damages to a variety of structures, the Seismic Safety Commission attributed much of it to lack of compliance with building codes and to low quality design and construction of structures. 'Continued economic pressures to lower design and construction costs have eroded the quality of both design and construction. Low-cost projects, which include both low-cost designs and minimum-quality materials, were present in significant failures in the Northridge earthquake' (SSC 1995: 23). The earthquake did cause significant damage to at least two dozen steel framed buildings, previously thought to be the most seismically resistant forms of construction. Damage occurred in welds holding steel spans together where cracks formed, although there were no actual steel frame building collapses. Because the damage to the frames is difficult to detect several buildings were red tagged more than two years after the main shock as detailed inspections revealed extensive cracking. Repairs to these structures is very expensive and the seismic safety of repaired structures is an engineering unknown. This phenomenon has raised numerous concerns about the performance of such buildings in 'the Big One' and whether other buildings have hidden damages from earlier earthquakes heretofore undetected (SSC 1995).

Managing an emergency

With its hypocenter under the most populous county in the US, the earthquake produced a large multi-organizational emergency response operation (e.g. Nigg 1997). As emergency management in the US is the direct responsibility of local authorities (city and county), the Northridge response was handled by a large intergovernmental effort of local and county agencies and departments (police, fire, emergency medical services, building safety, transportation). The state of California, through its Office of Emergency Services, played a central coordination role in the Northridge disaster. OES has a well established and ongoing presence in Southern California due to the numerous disasters the area has experienced in the 1990s. The federal government, under the lead role of FEMA, entered the Northridge disaster at the request of the governor of California on the morning of the earthquake. Both OES and FEMA provided substantial financial, material, and organizational resources to affected parts of the disaster area, using a total combined staff of 5,000 at the height of management activities.

The US government has been developing a federal response plan (FRP)[6] for a number of years. The plan is designed to coordinate the disaster response activities of twenty-six federal agencies and the American Red Cross. The ARC has primary responsibility for the immediate shelter and material support needs of persons displaced by disaster. Following the presidential disaster declaration on 17 January 1994, the FRP was activated. The core of the program is the division of federal agency activities into twelve different emergency support functions including emergency sheltering, health care, logistics, information and planning, and hazardous materials problems. Paralleling the federal mobilization, the state of California activated its own emergency response. The state has been implementing a Standardized Emergency Management System (SEMS) in order to coordinate activities among the numerous state agencies involved in disaster management (Nigg 1997). When fully operational, SEMS is designed to ensure that emergency management activities of a diversity of city and county authorities are consistent across the state. At the time of Northridge many California cities, including Los Angeles, were developing a SEMS framework for emergency management. Analogous to the federal response plan, SEMS organizes emergency management into functional areas including logistics, operations, administration/finance, and planning. The standardized emergency response plans facilitated rapid mobilization and coordination of personnel and resources in the area.

The array of dozens of federal and state agencies were internally coordinated at two different sites in the Los Angeles metropolitan region. The California Office of Emergency Services has a Regional Emergency Operations Center (REOC) located in the community of Los Alamitos in the County of Los Angeles. Immediately after the impact this center was mobilized by OES and became the center for initial coordination of the numerous federal, state, and non-profit agencies that began converging. The REOC was the managing center for emergency requests from local authorities, where staff attempted to match incoming resources with local requests for assistance as the regional emergency unfolded. OES sent representatives to each county's Emergency

Operations Center to relay county requests for assistance to the Los Alamitos center. A number of agencies at the federal level, including FEMA, Housing and Urban Development (HUD), Department of Transportation (DOT), and Army Corps of Engineers were present at the REOC. They joined representatives from California Highway Patrol, California Department of Transportation (CALTRANS), California Conservation Corps, the state Emergency Medical Services Authority, and the Red Cross and Salvation Army, to begin the response (Tierney 1996).

The second major center, and what became the main point of coordination for federal and state relief and recovery operations, was the Disaster Field Office (DFO), located in Pasadena, southeast of the epicentral region. Both FEMA and OES had ongoing disaster operations in Pasadena after the 'civil disturbance' in the South Central district (1992) and the Alta Dena firestorm (1993) in the foothills above Pasadena. The existing facilities became the base of the new DFO and were rapidly expanded within a few days of the main shock to accommodate the influx of federal and state disaster managers. On 19 January, the DFO became FEMA's operational center for the Northridge disaster. The DFO, at its height of operation, was a large and complex assemblage of federal and state agencies spread across several office buildings in downtown Pasadena. The physical juxtaposition of state and federal agencies in the same office complex facilitated interagency coordination as daily joint meetings and briefings could be arranged. Los Angeles was the only city, out of the dozens affected, that had representatives at the DFO, providing it with immediate access to both information and federal resources as the emergency developed (Tyler 1997b). The DFO was well equipped with data processing and mapping capabilities through the Geographic Information Systems Division (GIS) of OES. Agencies used information technologies extensively, making Northridge an extensively documented disaster, particularly in the areas of seismic and ground motion data, building damage data, and assistance program data (OES 1995, 1997).[7] The ongoing GIS mapping and data processing were a key part of managing and co-ordinating response and recovery activities across the complex political and physical landscape of the impact region. The most innovative information technology used a GIS-based system to project losses prior to actual information on damages being obtained. The Early Post-Earthquake Damage Assessment Tool, under development for OES, uses 'real time' seismic data, applying various ground motion models to provide loss estimates (buildings, lifelines, and casualties) based on mapped data of buildings, lifelines, and demographic information for a given area (Eguchi et al. 1997).

The key emergency management activities that fell to local agencies were fire fighting, police functions, traffic control, emergency medical assistance, restoration of public utilities, and, in some instances, search and rescue. It is well recognized that it was the victims themselves that were the first responders for search and rescue, evacuations, and giving medical assistance in the immediate period after the main shock. The Los Angeles Fire Department has, for a number of years, trained community volunteers as part of its Community Emergency Response Teams program. Organized at the neighborhood level, volunteers are trained in basic search and rescue and related emergency services activities. A similar program was in place in Ventura through the Red Cross volunteer Disaster Action Teams. People trained in both programs were active in the early hours of the disaster, pending the arrival of official emergency services personnel. Major power outages and the resultant darkness, collapsed highways, ruptured water mains, and interruptions in telephone service, all conspired to produce inevitable delays in the dispatching of fire and police units to areas needing assistance. Because it sustained the most concentrated damage, the greatest demands for emergency assistance were in the San Fernando Valley (Nigg 1997).

Northridge itself was the site of the worst single building collapse when the first story in a three-story wood framed apartment failed killing sixteen occupants. Because another thirty persons were trapped in the collapsed first floor of the apartment structure, it became the center of a large search and rescue operation. In all it involved seven Los Angeles City Fire Department units as well as four Urban Search and Rescue Task Forces. The US&R teams are supported by the OES as part of the state's overall emergency management strategy. A collapsed parking structure in a Northridge shopping center was also the site of a heavily televised rescue of a single worker trapped between levels of the structure. The rescue involved a significant commitment of personnel and equipment. While there were other search and rescue activities, these two incidents were the most time-consuming and demanding for emergency services personnel. The OES-authorized use of the US&R teams is generally credited with freeing up fire department units to combat the numerous natural gas fires that were ignited after the main shock.

High demand for fire suppression and related emergency services was met through California's elaborate mutual aid structure. The state itself is divided into six OES Mutual Aid Regions. Once a local emergency management operation determines that it needs additional assistance, then the mutual aid system is activated and personnel from elsewhere in that region are dispatched to the areas in need. OES staff coordinate and support the mutual aid assistance that comes into the local area. The counties of Los Angeles and Ventura are part of the same mutual aid area (Region I), as are Orange, Santa Barbara, and San Luis Obisbo counties. The mutual aid system was used to support emergency response in the San Fernando Valley, police and fire activities in Ventura County, and in damage assessments throughout the two-county area.

Fires in the San Fernando Valley were the result of ruptured natural gas mains which were ignited causing fires that spread to nearby structures. In addition, there were several major fires in mobile home parks in the valley as homes were shifted off their foundations, gas lines were severed and fires were started. In three separate fires a total of 184 mobile homes were destroyed, although no one was killed in the conflagrations. The other major fire was in Pacoima, near the city of San Fernando in the eastern San Fernando Valley where a large oil pipeline ruptured spilling oil onto a surface street which then ignited burning down two homes. The fifty separate earthquake-related fires in Los Angeles County were under control within a few hours of the initial main shock.

The power outages and their attendant affects on computer systems did create problems and delays in the dispatching of emergency services in both Los Angeles and Ventura Counties. Medical services were likewise constrained by the loss of water and power to facilities, in addition to damages to thirty hospitals across the area. The state Emergency Medical Services Authority, as part of its mutual aid coordination function, placed eight Disaster Medical Assistance Teams on alert following the emergency declaration. DMAT teams, at the request of the county were sent to the San Fernando Valley, Santa Monica, Canoga Park, and Valencia (Santa Clarita) to assist with medical care. However, with approximately 9,000 earthquake injuries, the health care system was not heavily stressed. The other major medical demand came in providing health care for a refugee population in excess of 20,000 living in Red Cross shelters and 'tent cities' erected in parks across Los Angeles County.

In addition to the various emergency services functions, damage assessment constituted a major challenge for local governments. Because of widespread damage, and the very large number of buildings subjected to ground motion in Los Angeles and Ventura counties, city and county building inspectors simply could not handle the demands in a timely fashion without additional inspectors. Although the City of Los Angeles has a staff of 500 building inspectors in its Department and Building and Safety, there were more than 75,000 structures that needed rapid inspection to determine structural integrity. Quick and accurate inspections are of key importance, both in determining if residents can return to buildings (i.e. green tagged) and in providing preliminary damage estimates to the federal government so assistance programs can be funded at appropriate levels. The mutual aid system brought in hundreds of building inspectors to assist local authorities in the two-county area. Part of the assistance came in the form of 280 inspectors on loan from the Army Corps of Engineers, in addition to more than 500 others from around the state provided through agreements with five professional engineering and construction organizations. The placement of colored placards in buildings is an important part of emergency management as it alerts both occupants and responders as to the relative safety of structures.

As the City of Los Angeles sustained the greatest damage to buildings and highways, it was of a primary focus of attention from the media as well as the federal government in the immediate aftermath. It can also be said with reasonable confidence that both the City and County of Los Angeles are better prepared and organized than any other metropolitan region in the United States to handle a disaster the size of Northridge (e.g. Nigg 1997). This preparedness and management capacity is the product of experience in handling seven federal disaster declarations in the five years leading up to Northridge (and two since the earthquake). Given the national significance and influence of both Los Angeles and the state of California, the disaster quickly became a high profile political occasion and a focus of political claims-making activities at both the state and federal level.

The politics of crises

The Northridge disaster emerged as a focal point of political discourse and it became an arena in which claims of disaster needs were publicly promoted at both the state and national levels. The political 'weight' of Northridge was reflected in the character and scope of the federal response. To begin with, the immediate convergence of leading members of the Clinton administration insured the political construction of Northridge as a major disaster and created the political base for a sustained federal presence for relief and reconstruction assistance. The director of FEMA, James Lee Witt, Secretary of Transportation, Federico Pena, and Secretary of Housing and Urban Development (HUD), Henry Cisneros, all arrived in Los Angeles by evening of the first day. After having declared the Los Angeles region to be in a state of emergency, Governor of California, Pete Wilson, and OES Director, Richard Andrews, were in LA within hours of the main shock. FEMA had personnel on site in Los Angeles early that morning and had dispatched materiel and disaster teams before a presidential disaster declaration had been issued on the afternoon of 17 January 1994. The immediate presence of Cabinet members in addition to top-ranking California politicians in a heavily televised and publicized event lent an immediate political momentum to Northridge that would become instrumental in the passage of a large but politically contested assistance package to Southern California.

One of the political subtexts of the disaster was that it arrayed a conservative Republican state governor (Wilson) against a first term Democratic president (Clinton) and a new director of FEMA (Witt). Clinton was also the first Democratic president to carry California in national elections in twenty-eight years, underlining to the political salience of the state for his administration and for his re-election bid in 1996 (*New York Times*, 19 January 1994). The fact that the earthquake occurred during a congressional election year in a state with the largest electoral vote in national elections added to the political weight of Northridge. Thus, the immediate presence of Cabinet level officials, followed in a few days by a visit from president Clinton, and then a subsequent visit by the president's wife, were all elements in the political construction of Northridge as a high profile disaster with a major federal presence. The fact that these early offers of assistance were made on national television lent a near irreversibility to the commitments. Transportation Department Secretary Pena, on 19 January, committed to 100 per cent federal coverage of repairs to interstate highways for six months, without the usual requirement of 25 per cent state matching funds. HUD Secretary Cisneros likewise ordered the immediate disbursement of $129 million in Community Development Block Grants (CDBGs) to the region (see Chapter 5), in addition to the provision of 20,000 Section 8 housing certificates for low income victims needing housing. All these offers of federal support were made well in advance of all but the most preliminary loss estimates.

Damage estimates on 19 January were being published in the *LA Times* as approaching $7 billion, but would within days rise by 400 per cent. At the time of Northridge, FEMA had $1 billion in reserve funds left over from the massive 1993 Midwest floods. According to a FEMA spokesperson it would take several weeks from the time of the earthquake before FEMA would know if these reserves would be adequate to cover Northridge. Those funds would prove to cover only a fraction of federal obligation to Southern California by the time all programs were closed out two years later. Once public political commitments had been made for resources, even in the absence of any clear idea of what those commitments might mean, then disaster assistance became an important element of political legitimacy of the administration and failure to deliver the assistance would have been costly in terms of political capital lost.

The rapid mobilization at the federal level was facilitated in part because the governor of California declared a major disaster area and requested federal assistance within a few hours of the actual main shock. FEMA, which had been sharply criticized for a delay of several days before providing assistance after Hurricane Andrew (Florida) in 1992, significantly improved their entire response capabilities and began providing assistance for Los Angeles even prior to the federal declaration that authorized it. The fact that Los Angeles County had already received a number of recent federal disaster declarations also meant that both state and local officials were familiar with federal policies and procedures and could rapidly build a response operation in close coordination with FEMA.

On 12 February, after going through a series of congressional hearings, an initial $8.6 billion earthquake aid bill[8] was signed by the president. For major disasters supplemental appropriations to cover federal commitments have to be proposed by the president and submitted for approval by the congress. A political stalemate ensued as conservatives attempted to fund Northridge from monies taken from other already stressed social programs, rather than deficit spend. The political contestation delayed passage for several days as Republicans attempted to force yet further cuts in social spending. Conservatives successfully attached a rider on the appropriation requiring that FEMA determine if applicants seeking assistance were 'lawfully within the US', subject to verification by the Immigration and Naturalization Service. This applied to anyone applying for FEMA assistance beyond a ninety-day period and is emblematic of a broader political thrust to deny social benefits to undocumented immigrants. Demand for federal aid was high in spite of exclusionary statutes and by mid-February more than 300,000 applications for federal and state assistance had been received (Chapter 5).

State politics surrounding financing Northridge losses were also not without conflict. A $1 billion state financial obligation through OES, CAL-TRANS, and related operations seemed nearly inevitable, even with generous federal cost sharing arrangements. At the same time a persistently stagnant state economy and weak state revenues had produced budget constraints in numerous state programs. Because many federal programs require matching funds from the state, usually at a 25–75 per cent state–federal ratio, disaster programs such as Individual and Family Grants (IFG), highway construction, public building repair, and school rebuilding all could prove very costly to the state and local governments. The governor made a request on 20 January that the federal government commit to providing 100 per cent coverage of reconstruction costs for public buildings and highways, noting that estimated costs of the disaster were between $15 and $30 billion. Part of Wilson's claim was that California was already spending 'billions of dollars' on illegal immigrants for various social services and consequently deserved a large federal disaster reimbursement (*Washington Post*, 26 January 1994). The federal government took into account the state's weak economy and its numerous recent disasters and granted a 10 per cent state match on federal assistance.

The major early political debate within the California legislature, conducted while the extent of damage was still being determined, was over what fiscal mechanism to employ to raise revenues to pay for the state share of disaster programs. After the $6 billion Loma Prieta earthquake in 1989, the state levied a small increase in sales tax to provide revenues for state obligations for recovery. Democrats in the state legislature supported a similar strategy to raise the estimated $750 million or more necessary for Northridge. However, the governor opposed any tax increases as a revenue source since tax increases are notoriously unpopular, particularly for a conservative incumbent facing election in a state that had popularized 'tax revolts' (e.g. Lo 1990). The alternative financing mechanism proposed by the governor was the sale of state bonds to cover earthquake costs, including an extra $1 billion for highway construction and retrofitting projects. Neither the state sales tax increase nor the bond measure were passed, leaving the state with no special funding mechanism to cover its obligations.

Thus, far from being a 'consensus event', the Northridge disaster engendered immediate political splits over long-term financing and government (federal and state) obligations to both the public and private sector (cf.

Olson and Drury 1997). Political contestation began within days of the main shock and tended to attach unrelated issues such as 'illegal immigrants' to the funding debate at the federal level. (The limitations on federal assistance to those without legal residency was extended in 1996 to deny any federal disaster assistance to nonresidents.) Equally significant, given the frequency of its disasters, California's approach to financing significant disaster response and recovery operations was (and is) largely *ad hoc*, and subject to occasionally time-consuming legislative proposal and counterproposal. Politicizing the crisis and contesting recovery funding has specific consequences for the resources that are made available and for whom they are made available. Programs for recovery and reconstruction and the issues surrounding their accessibility are examined in Chapters 5 and 6.

Chronology of a disaster

The Northridge earthquake occurred in an urban region that is generally well prepared to manage a disaster. The area also has, as we have noted, some of most rigorous seismic safety standards in North America. In addition, as a result of the numerous disasters that have been declared in Southern California, it is an urban region that has emergency preparedness organizations well rehearsed in dealing with environmental catastrophes. Nevertheless, managing the disaster that emerged from Northridge proved a significant challenge to the emergency response system of the region as well as to longer-term recovery management. Recovery issues, which are discussed in Chapter 5, have been directly affected by the class and ethnic diversity of the area and its existing social, cultural, and economic problems described above.

There was an immediate federal response to the governor's request for assistance and FEMA officials began arriving in Los Angeles on 19 January for damage assessment and determining the scope of federal assistance warranted. As damage reports became available from Ventura County to the north and Orange County to the south, they too were included in the disaster declaration area on 19 January. While Ventura County had heavy structural losses in three communities, Orange County had in fact received very little damage. The late inclusion of Ventura in the disaster declaration gave rise to complaints regarding its peripheral importance in the federal response effort, a point raised again in the following chapters.

On the day of the disaster, the American Red Cross established twenty emergency shelters and the Salvation Army set up another four in Los Angeles and Ventura counties. Following the main shock thousands of people moved out of structures and into parks and other open areas out of concern over building safety during aftershocks. Los Angeles officials estimated that as many as 20,000 people were camped in parks, front yards, or parking lots, a pattern very similar to previous urban earthquakes in California. In general, there was a demand for shelters far in excess of capacity on the first day of the disaster (OES 1997). The Los Angeles City Department of Parks and Recreation is responsible for managing parks as emergency shelters during disasters and began immediately after the earthquake to assist those camped in parks. They managed more than thirty transitional shelter sites in parks and parking lots to assist evacuees until the ARC and Salvation Army had shelters fully operational. All transitional sites were closed down by 29 January, as victims either moved to ARC shelters or found other accommodations (Figure 3.5).

In addition to these informal transitional sites several larger refugee centers in parks were established, with tents provided by the California National Guard. Many who fled their homes, particularly Latino immigrants with recent earthquake experience, preferred tents to school auditoria and other buildings used by the ARC for shelters. The onset of major winter storms in late January made tent facilities increasingly uncomfortable both from rain and cold temperatures. Shelter populations peaked one week into the disaster and then declined steadily. The last park shelter was closed down on 19 February. The majority of those using Recreation and Parks facilities were Latino according to departmental surveys and nearly half were from green tagged buildings that they were fearful to return to (OES 1997).

In addition to those sheltered in the park facilities, the ARC provided shelter to 28,500 during their operations in the two counties in a total of forty-five shelters. (This number may include some number of the 20,000 sheltered first in parks by the City of Los Angeles). The peak shelter population housed by the Red Cross was 17,500 on 19 January. There was steady turnover in this population as people returned to homes and others had to leave as buildings were condemned after being inspected. As with park shelters, ARC shelters housed a combination of disaster victims and the pre-disaster homeless. According to a Red Cross survey, at least 13 per cent of shelter occupants had been homeless prior to the earthquake, a phenomenon also observed after other recent California earthquakes (Phillips 1993). Although as many as 88,000 people left their damaged housing units, only a small portion actually used official public shelter in Los Angeles or Ventura counties; most went to the homes of family or friends, or to commercial rental units. Surveys after Northridge indicate that the majority of shelter occupants were lower income persons who lived in rental units (OES 1997).

By 19 January there was a significant federal presence in the region as various agencies began assembling staffs and organizing the assistance programs. FEMA's telephone registration system for assistance applications was operative, although inadequate, at this time. The following day eleven Disaster Applications Centers were opened, nine located in Los Angeles County and two in Ventura County. At least one new DAC opened on each of the next six days across the two-county disaster area. By 8 February a total of twenty-one DACs were in operation in addition to fourteen mobile DACs

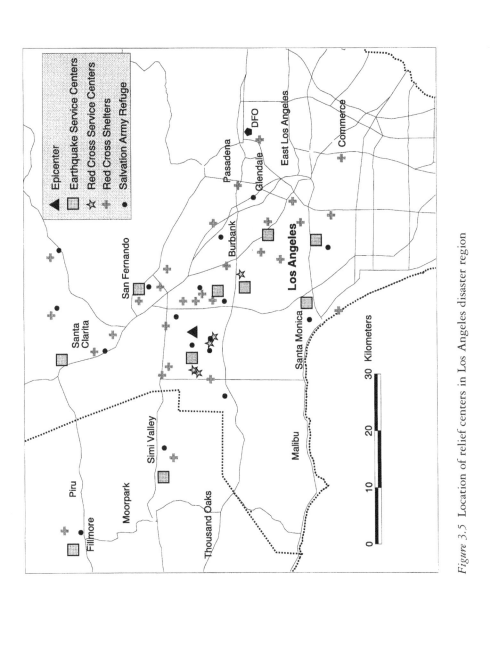

Figure 3.5 Location of relief centers in Los Angeles disaster region

that could respond to specific area demands. Another six DACs specializing in serving businesses damaged in the earthquake were also opened.

The Disaster Application Centers began to be phased out of operation beginning on 17 February. This did not signal an end to a direct federal presence, however. Earthquake Service Centers (ESCs) were established to take over services provided by the DACs in the two-county disaster area (Figure 3.5). The original twenty-one DAC's services were consolidated in ten ESCs during the second month of federal operations. The ESCs were part of a new federal and state approach which provided a long-term community presence for both OES and FEMA, giving victims a location to both apply for assistance and check on existing applications. At least some Earthquake Service Centers remained in operation for two years after the earthquake.

Due to a sustained high application rate for individual assistance after the disaster, FEMA repeatedly extended deadlines for each of its various programs. The original deadline for all individual assistance, 17 May 1994, was reset on several occasions, usually at the request of the governor, and ultimately set to terminate on the first anniversary of the earthquake. Applications received after the cut-off date were still considered on a case-by-case basis and processed if extenuating circumstances warranted it, an unusual concession by FEMA. There was a sustained application rate for some forms of assistance throughout the first year of the disaster, encouraged both by the ongoing presence of responders and active outreach programs through the ESCs. Many applicants entered the aid system at a comparatively late date because they discovered hidden damages on their residences they could not afford to repair. Others received inadequate settlements from insurers and turned to federal assistance to help cover differences between actual and insured losses. Thus as the close of applications neared, 500 applications a week were still being processed by FEMA, an indicator of the lengthy nature of earthquake recovery (OES 1997).

Given the critical nature of the region's vast freeway infrastructure, the state, with federal highway funds, moved quickly toward restoration. Various traffic management strategies were developed in the first month of the disaster to deal with lengthy traffic delays from the loss of highway capacity. These strategies included routing traffic off freeways on to city streets, adding bus services, and creating special traffic lanes for people who would car pool to work sites. To expedite repairs of the highway system, the governor ordered the suspension of normal state contracting procedures and contractors were hired by the California Department of Transportation (CALTRANS) to begin demolition of damaged highways within days of the earthquake. On 31 January, the state issued its first highway repair contract and other contracts quickly followed. Each contract specified the number of days to repair a section of highway and contractors were offered a daily bonus of $150,000–$200,000 for every day they finished ahead of schedule. The damaged section of Interstate 10 was finished seventy-four days ahead of schedule earning the contractor nearly $15 million in bonuses. Interstate 5 connecting Santa Clarita to the rest of Los Angeles County was completed one month ahead of schedule on 17 May, garnering its contractor additional millions in bonuses. Although it cost the state more than $20 million in early completion bonuses, most major freeway sections were repaired in five months, allowing traffic to begin to return to 'normal'.

At a point four months into the disaster, federal and state agencies had committed or spent $4.1 billion, including individual and business grants and loans, as well as federal assistance to local governments for public infrastructure repair. Of this, nearly $1 billion had gone for temporary housing and emergency home repairs. Another $1.6 billion had been approved in SBA loans to individuals and businesses. For residential units requiring extensive repairs, the process could be drawn out over a year or more depending both on when assistance monies were received and on the speed at which repair work was actually carried out. For homeowners with earthquake insurance, the time to settle claims could be lengthy, routinely measured in months. If the settlement was disputed by the policy holder, a relatively common occurrence, the time period required to obtain a settlement could be more than two years. Nevertheless more than 265,000 claims were filed by homeowners and private insurers proved to have an unexpectedly high liability exposure in Northridge. Payments on homeowner claims are now placed in excess of $8 billion and the total insurance payout exceeds $12 billion (Comerio et al. 1996; OES 1997).

By July 1994, six months after the disaster began, federal financial commitments to the disaster region had risen to $5.5 billion. By this time, due to additional releases of federal funds, the total federal allocation was $11.2 billion, and would increase another $1 billion with a FEMA grant to repair a major public hospital in Los Angeles. In addition to more than $1 billion for temporary housing, FEMA had also disbursed $714 million in public assistance to local governments to assist them in repair of infrastructure and public buildings. SBA loans had been approved for $2.7 billion and HUD assistance had exceeded $625 million for rent subsidies and Community Development Block Grants (see Chapter 5). The Federal Highway Administration had authorized payments of nearly $350 million to repair the region's crumpled freeways and bridges, many of which were back into operation before six months had passed.

Northridge in context

Discussed in aggregate terms, Northridge may appear as an extreme event. Yet when such numbers are considered in the context of a conurbation the size of the Los Angeles region, it is not. After all, it is a First World 'megacity' with more than ten million inhabitants in a state with a gross domestic product of $735 billion (Hundley 1992). The City of Los Angeles alone has a staff of 35,000 to manage its day-to-day operations (Tyler 1997b). The large number of applicants for federal and state assistance reflects the dispersed effects of the earthquake across this large resident population. They also may be read as a sign of an atypical federal willingness to keep application periods open for long periods and to engage in an outreach effort to encourage applications. Overall, a concerted and costly effort was made in the North-ridge event by lead federal and state agencies to restore the Los Angeles area's large-scale infrastructure to more or less routine functioning and, where feasible, to secure structures against future seismic events.

Many of the specific response and recovery issues that we address in the following chapters relate to historical trends that have shaped the region's economy, its demographic composition, and its urban form. The sheer scale of the region combined with its complex network of lifelines and its equally complex web of political jurisdictions mandate a large well-organized emergency response system to coordinate critical response activities. The sociocultural elements that have made the region an archetypal postmodern landscape (Baudrillard 1988; Dear 1996) also affect both the general characteristics and the local manifestations of a disaster like Northridge. The polyethnic mix of cultures and nationalities combined with extreme income inequalities produce different forms of vulnerability across the area. Imbricated in this are politically marginalized immigrants who lack legal residency and who were intentionally excluded from federal disaster assistance as a consequence. Yet on the other side of this class and ethnic divide there are significant numbers of middle-class homeowners who have lost equity in their homes as real estate prices have declined significantly since the 'boom' days of the mid-1980s (Comerio 1995; Davis 1992). They too can be vulnerable to a common disaster like Northridge. And among the shrinking middle class, there are some who have lost jobs from cutbacks in the aerospace industry and its once lucrative military contracts (e.g. Soja 1996b).

In sum, the Northridge earthquake set into motion numerous efforts on the part of victims, government agencies, and communities to manage the crisis and recover from it. The earthquake prompted major federal and state responses intent on stabilizing the region, minimizing economic disruptions to the production system, and to providing assistance for individual householders. The federal government has maintained a major interest in supporting repairs to both public infrastructure and businesses with an extensive program of loans and grants. However, despite all the money spent, state supported efforts to repair structures and reduce future earthquake losses in the built environment rely largely on voluntary compliance and imple-mentation; as part of the *realpolitik* of disaster, they do little that might threaten growth or profits in the name of earthquake safety.

The summaries of the losses and assistance are one part of understanding Northridge. Yet this 'view from above', as familiar as it may be as a way of describing disaster, gives us only the panoptic view of the government's and private sector response system and its efforts at managing a decentralized disaster. There is no single grand narrative of the Northridge disaster, and although the aggregate statistics may give some impression of the totality, they fail to capture the local features of the disaster as experienced in the day-to-day lives of people and their communities. The complex, multilayered, and fragmentary nature of the social and geographical landscapes of the Los Angeles region defy easy summarizing as do the experiences of the people who live there. As we will begin to develop in the following chapters, historical and geographic trajectories of the region connect in various ways to the localized conditions of communities affected by Northridge. The historical features sketched out here are a background against which to examine the elements of vulnerability at specific sites. It is at the local level that people's vulnerabilities become most easily visible, and it is at the local level that strategies to recover and to address vulnerability manifest themselves most concretely.

Our attention in subsequent chapters turns toward the 'peripheral valleys' (Soja 1989) of northern Los Angeles County and Ventura County, where we consider four communities as case studies. These illustrate the mix of response and recovery issues, and the ways local strategies were developed to address recovery and the problems of vulnerable households. As we discuss the specific experiences of this common disaster in the four communities, we also compare local developments to regional recovery issues.

NOTES

1 It was not the lack of water that allowed the fires to burn, but the inability to deliver the water for fire fighting due to ruptured water mains (Hundley 1992). While water for fire fighting may have been the justification for the expropriation of Yosemite water, little was done by the State of California regarding earthquake safety after the San Francisco earthquake. By any account, the occasional earthquake was simply not a concern of the promoters and developers of the state (e.g. Davis 1992).

2 The plotting of new towns around the core of the City of Los Angeles during this time was the beginning of a process of spatial dispersal that continues to the present. As Soja (1989) notes, Los Angeles bypassed the centralization of most nineteenth-century industrial capitalist cities, driven by real estate speculation more than by industrialization. The metropolitan region of Los Angeles includes some or all of Los Angeles, Ventura, Orange, San Bernardino, and Riverside counties. It is a complex political mosaic of city and county governments with jurisdictions over a range of municipalities, and unincorporated areas. In addition to the City of Los Angeles, Los Angeles County houses more than eighty other cities, each with its own city government, as well as 145 named districts and 'minimal cities' (Miller 1981) that contract most of their urban services from the county of Los Angeles. According to Scott and Soja (1996) the Los Angeles metropolitan region is the sixth largest 'megacity' in the world. The term Southern

California is used to designate an area inclusive of the entire Los Angeles metropolitan area in addition to the San Diego metropolitan area to its south, and has a population of fifteen million (Davis 1992).

3 NEHRP has recently been joined by the National Earthquake Loss Reduction Program (NEP). Whereas NEHRP involves four federal agencies, FEMA, USGS, National Institute of Standards and Technology, and the National Science Foundation, the NEP is a broader effort at coordinating some fifteen federal agencies also involved in earthquake hazard activities. As with NEHRP, the NEP lead agency is FEMA and a strategic plan for earthquake hazard actions is currently being formulated.

4 The South Central district is one of the City of Los Angeles' six Community Improvement Planning Areas. It is a largely poor multi-ethnic district that was the scene of major urban riots in 1965 and 1992 (Soja 1996b). The 1992 riot followed a televised court trial in which Los Angeles police were accused of beating a black man for a traffic violation. The beating was videotaped and played repeatedly on news programs. The trial concluded with the exoneration of the police officers and a major 'civil disturbance' followed resulting in considerable damage to buildings and businesses in the district. The district was declared a disaster area and FEMA began recovery operations. The Northridge disaster further damaged this area with its older, often poorly maintained housing stock.

5 Buildings are listed in reports as being red, yellow or green tagged, in reference to the color of placards placed in them by building inspectors to indicate their relative safety. Red tagged buildings are unsafe for any but official entry. Yellow tags indicate a building has sustained major damage but is safe for limited entry. Green tags indicate that the building is safe for occupancy although it does not mean that the structure is undamaged. A green tagged house may have suffered thousands of dollars of 'cosmetic' damage from ruptured water lines, falling light fixtures, broken windows and cracked plaster.

6 The scope of the response plan has been recently broadened and is now called the Federal Response and Recovery Plan, largely as a consequence of the Northridge disaster operation.

7 While OES and FEMA have made some of their data sets available to researchers, confidentiality restrictions limit the kinds of information that can be released. This is particularly true of Individual Assistance Data, which contains no information on household characteristics of applicants by income, ethnicity, or gender. This limits the utility of these data for social analysis beyond referring to aggregate census data against IA data for the same census tracts (e.g. Comerio et al. 1996).

8 Funding for major urban disasters in the US is done on an event-by-event basis, depending on the estimated losses and other determinations of the federal role in assistance. An $8.6 billion supplemental appropriation was passed by the US congress and signed on 12 February 1994. Additional federal funds brought the total disbursal to more than $11 billion. As $4 billion of this is through SBA loans, that portion will ultimately be paid back with interest.

5

RESPONDING TO
NORTHRIDGE

The Northridge earthquake triggered a complex set of organizational, household, and individual actions across a large and dispersed urban area. In Chapters 3 and 4 we considered historical factors that have shaped the urban forms and structures of social inequality across the region. These provide contexts for examining the patterns of vulnerability revealed by the earthquake and the social actions engendered in its aftermath. In this chapter and the next we partition the discussion of the Northridge disaster into two frames. Here we discuss general and localized responses to the disaster, specifically focusing on individual losses and the programs intended to assist victims. The term 'responding' is intended to connote the heterogeneous and processual nature of activities undertaken by people, organizations, and governments to cope with the disaster and recover from its direct effects. We avoid analytically dividing the disaster into discrete 'phases' (emergency response, restoration, recovery), since such partitions – while having some heuristic value – tend to gloss the complex and contradictory ways different actions can combine in uneven temporal sequences (e.g. Neal 1997).

In the next chapter, we discuss programs which, while initiated as part of recovery, extend into the realms of social change and development. Conceptually, actions that restore conditions to what they were before the earthquake are considered 'recovery', while those that involve organizational and economic changes are termed 'restructuring' (e.g. Albala-Bertrand 1993). Some social actions and programs involve both simultaneously as with City of Los Angeles' efforts to promote restoration in heavily damaged neighborhoods in the San Fernando Valley (Chapter 6). Using this conceptual distinction, actions that seek to reduce vulnerability will often involve changes in structural arrangements in communities, hence are considered part of restructuring. Economic development, social change, and the structural alterations these involve are ongoing processes. In a common disaster such as Northridge, the event may be of only incidental significance to longer-term trajectories, a means of highlighting existing problems while augmenting resource availability and creating new political opportunities.

In California, existing organizational capacities and intergovernmental networks provide high levels of social protection against disasters, although programs and the extent of protection vary among municipalities and socio-economic groups. Federal and state actions after the earthquake, activated as part of a national infrastructure of disaster management, moderated the disaster's impacts through a variety of shelter, temporary housing, and recovery programs. In this chapter, we first review relief programs that were generally available across the entire disaster area, and we examine routinized aspects of response and recovery, specifically federal programs directed towards housing support and rehabilitation. Our focus narrows as we consider local response and recovery issues that emerged in the study sites. A consideration of local conditions and responses reveals dimensions of household vulnerability and the emergence of programs intended to address those with unmet recovery needs. The conclusion returns to the conceptual framework of vulnerability introduced in Chapter 2, and reconsiders it in light of these local contexts.

INSTITUTIONALIZED DISASTER RELIEF

Institutionalized disaster relief – programs routinely available after federally declared disasters – includes federal and state programs, emergency relief services by the American Red Cross (ARC), and private hazard insurance. The major government assistance programs widely available to individuals and households after Northridge are listed and briefly described in Table 5.1, and estimates of total federal and state expenditures are summarized in Table 5.2. These federal programs constituted about half of all monetized direct expenditures in the Northridge disaster. (There is no accurate way to determine total uncompensated expenditures made by households that used personal resources to pay for earthquake damage.)

The federal government responded to the earthquake with a $12 billion aid package delivered through an elaborate system of service centers, supplemented by additional monies made available by the state of California. A total of 681,117 applications for individual assistance (IA)[1] were received in the one-year period that they were accepted (ending 20 January 1995). This one-year period established a new maximum for FEMA, surpassing the 7.5 month period for the 1989 Loma Prieta earthquake in Northern California. (FEMA accepted late applications for temporary housing assistance on a case by case basis for almost a full year beyond the initial cut-off.) The number of applicants for individual assistance almost doubles the previous record, set after Hurricane Hugo struck the southeast Atlantic coast of the US in 1989 (FEMA 1995). While the number of relief applications gives some idea of the scale of the disaster, that number underestimates the affected populations. OES (1997) estimates that more than 6.6 million people in the tri-county disaster area were exposed to a Modified Mercalli Intensity (MMI)[2] shaking of VI or higher, suggesting that a very large number of households experienced

Table 5.1 Major individual assistance programs available after Northridge

Agency	Program	Service
FEMA	Temporary housing	Grants for rental assistance – 3 months for owners, 2 months for renters, $1,150 per month – owners can be extended to 18 months
	Mortgage assistance	Mortgage payments for disaster unemployed homeowners – up to 18 months
	Minimum home repair	Grants to repair minimally damaged houses – up to $10,000
SBA	Real property	Low interest loans to repair homes – up to $200,000
	Personal property	Low interest loans to replace home contents – up to $40,000
OES	Individual and family grants	Grants to individuals with unmet recovery needs – $22,200 maximum, including special $10,000 supplement
	Mobile home repair	State managed program to repair and brace mobile homes on foundations
HUD	Section 8 housing vouchers	18 months rental assistance for very low income households
ARC	Mass care	Emergency shelter, food, clothing, psychological counseling
	Family services	Short-term housing and related assistance, referrals, advocacy
Salvation Army	Disaster services	Emergency shelter, food, clothing, social support

Notes
FEMA – Federal Emergency Management Agency
SBA   – Small Business Administration (federal)
OES   – California Office of Emergency Services
HUD – Department of Housing and Urban Development (federal)
ARC   – American Red Cross (NGO)

Table 5.2 Estimated total programmatic expenditures

Program area	Estimated costs
Small Business Administration (homeowner and business loans)	$4,030 million
Individual and family grants (federal expenditure including mental health program)	$243 million
FEMA housing programs	$1,200 million
Housing and Urban Development grants, etc.	$836 million
California programs and cost-sharing	$1,000 million
Public assistance (FEMA programs)	$4,500 million
Transportation infrastructure (federal)	$327 million
Utility repairs (federal)	$300 million
American Red Cross and Salvation Army	$37.25 million
Total expenditures by responding agencies	~$12,500 million

Source: OES (1997)

at least minor damage even if they did not seek assistance. The cost of residential reconstruction is placed at $12.6 billion, nearly half the total direct losses of $25.7 billion as calculated by Comerio *et al.* (1996).

Assistance programs and utilization

Private earthquake insurance was a major source of home rebuilding assistance after Northridge, paying out an estimated $7.8 billion in more than 265,000 residential claims by March, 1995 (OES 1997). Current estimates of earthquake insurance payouts to individuals and businesses have risen to $12 billion, making it the single largest source of reconstruction funds (Comerio *et al.* 1996), benefiting primarily middle and higher income homeowners. (The exception is the case of mobile homes, which, although relatively inexpensive, have earthquake insurance as part of their regular homeowner policies.) An estimated 35 per cent of the single-family dwellings affected by Northridge had earthquake insurance, nearly twice the rate of insurance coverage after the 1989 Loma Prieta disaster. Approximately two-thirds of all insurance payouts in Southern California were under homeowner policies, at an average of $29,000 per claim (Table 5.3). In contrast, 12 per cent of the claims were under condominium policies and 6 per cent under renters' policies, with average payouts of $7,300 and $5,200 respectively (Comerio *et al.* 1996; OES 1997). For the low income homeowners and renters at the research sites, few, other than mobile home residents, had earthquake insurance on homes or businesses. For vulnerable households in the four communities, federal, state, and NGO relief programs were the primary assistance sources.

In the immediate aftermath, several dozen emergency shelters and tent encampments were established in LA and Ventura counties and managed by the ARC and Salvation Army, with the aid of more than 15,000 volunteers. Most shelters remained open for three to four weeks, closing as demand decreased when victims returned to their homes or found new residences. The ARC's Family Services program also opened eighteen Service Centers where victims were offered small grants and housing vouchers to assist them until federal assistance was received. (Service Centers are different from DACs, although the two are frequently confused both by victims and the press.) The Service Centers were kept open for ten weeks and dealt with 33,700 cases.

FEMA's three Disaster Housing Programs provide financial assistance to people who cannot return to their residences due to damage. Under the Temporary Housing Program, renters whose apartments are uninhabitable can receive rental assistance for two months, with continuing assistance available on a month-by-month basis. Homeowners whose dwellings are uninhabitable receive an initial three months of rental assistance for temporary housing and may be eligible for up to eighteen months of support while homes are being rebuilt. Rental assistance after Northridge was disbursed to nearly 119,000 victims at a rate of $1,150 per month (a figure based on estimates of regional rental rates). FEMA also offers a Disaster Mortgage and Rental Assistance Program to those unemployed as a result of the disaster, allowing them to avoid foreclosure or eviction. (Those unemployed or already in arrears on mortgage payments at the time of the disaster are not eligible.) FEMA pays the actual amount of the mortgage or rent for up to eighteen months or until new employment is found, whichever is shorter. This program received 1,000 applications (FEMA 1995) reflecting the relatively limited effect of the earthquake on employment.

Homeowners whose dwellings sustain less than $10,000 in damages, as determined by FEMA inspectors, are eligible for grants from the agency's Minimum Home Repair Program (MHRP). These grants, issued shortly after a disaster, assist people in restoring their homes to a livable condition. After Northridge, MHRP grants, averaging $2,900, were given to 288,000 homeowners for a total of $841 million (FEMA 1996). Those receiving MHRP grants may also receive additional resources through other FEMA programs, although the amounts are generally modest in relation to regional home values. Eighteen months after the main shock (August 1995), FEMA had approved more than 407,000 applications for the three Disaster Housing Programs and disbursed nearly $1.2 billion in the three counties; 90 per cent of this went to applicants in Los Angeles County (OES 1997).

Victims who could not qualify for other federal assistance programs or who had unmet needs after receiving federal assistance could receive aid through the joint OES–FEMA Individual and Family Grant Program (IFG) (the state administers the program and pays 10 per cent of the costs). To qualify an applicant must have been turned down by the Small Business Administration

(SBA) for a loan and have failed to receive adequate assistance from other FEMA grant programs. IFG is normally used by those with limited resources and consequently is used by far more renters than homeowners. While IFG grants are normally used to replace personal possessions and home contents, they can be used for home repairs. For Northridge the maximum IFG payout was $12,200, with an additional $10,000 supplement available from the state of California for those with significant unmet needs. A total of 214,000 IFG grants were issued, with nearly two-thirds going to renters (OES 1997). However, homeowners received more than 90 per cent of the approximately 1,140 maximum awards. Those who received maximum IFG awards were often situationally vulnerable households: formerly secure middle-class homeowners who, due to a compounding of negative circumstances (illness, unemployment, uninsured earthquake damage), found themselves unable to qualify for SBA loans and lacked other means to pay for repairs.

Approximately $214 million in IFG funds were disbursed during the first eighteen months of the disaster, with more than $10 million going to applicants in Ventura County. The grants (which averaged less than $1,000) provided limited assistance for households that frequently had few resources to begin with. While IFG is the sole federal 'unmet needs program', there were numerous supplemental programs offered by NGOs and community-based organizations (CBOs) to assist those inadequately served by federal programs. The IFG program, as a program of 'last resort', is intended only to offer partial assistance and only for those with significant continuing deficits; it is not a substantial resource package that allows people to completely rebuild and recover. No federal disaster grant program is intended to do more than assist qualified victims to stabilize their lives in the short term.

Two other major sources of federal assistance to individuals were available after the Northridge earthquake. The SBA offered loans to individuals to pay for home reconstruction. These low interest loans cover non-insured losses on homes (real property), and on the contents of homes and apartments (personal property). For Northridge, the maximum loan was $200,000, with an additional $40,000 for personal property. (For comparative purposes, the median price of a San Fernando Valley home in 1994 was $230,000, rising to well over $400,000 in some Valley districts.) SBA also provides loans up to $1.5 million to businesses and rental property owners. Home loans are typically issued to middle or higher income owners with good credit ratings. The selectivity of the program is illustrated by the fact that only about half of the 194,000 applicants were able to receive loans. The 99,000 loans that were granted totaled nearly $2.5 billion (OES 1995). Two-thirds of those loans went to homeowners, 29 per cent to businesses, and the remaining 4 per cent to renters. The total SBA commitment to Northridge (including loans to businesses and individuals) is currently $4 billion, and it is the single largest source of federal assistance to individuals (OES 1997). According to Comerio et al.'s (1996) estimate, SBA loans accounted for 21 per cent of the total residential reconstruction funds available, and approximately 75 per cent of those loans went to residents within 20 kilometers of the epicenter. Those who used SBA loans also frequently received insurance payments suggesting that SBA loans were used to pay the large deductibles on homeowner's insurance policies. The average SBA loan amount is roughly comparable to average homeowner's insurance settlements (Table 5.3), but the loans require households to take on additional debt load for a period of years to pay it back with interest.

The Department of Housing and Urban Development also offered rental assistance and housing reconstruction grants to cities after the earthquake. HUD's emergency Section 8 Rental Certificate Program is offered to qualified households whose incomes are less than 50 per cent of median, providing rental assistance for a period of eighteen months. While not a normal part of federal disaster response, it was used after Hurricane Andrew in 1992 to assist a large low income victim population (Peacock et al. 1998). After Northridge a special $200 million block of Section 8 certificates was issued for very low income households displaced by the earthquake (FEMA 1995). This special disbursement was not without its critics, as earthquake victims received

Table 5.3 General distribution of rebuilding and repair funds by program

Source	Program/use	Amount (millions)	No. of recipients	Average award
Small Business Administration	Residential rebuilding loans	$2,500	98,847	$25,100
	Commercial rebuilding loans	$1,450	21,936	$66,100
FEMA	Individual and family grants	$214	214,277	$1000
	Minimum home repair grants	$841	287,778	$2,925
	Temporary housing grants	$381	119,583	$3,180
Private insurance companies	Residential repair/rebuilding	$7,808	265,116	$29,451
	Commercial repair/rebuilding	$3,400	15,708	$216,769
	Other coverages	$1,041	52,390	$19,882
Total		$17,635		

Sources: FEMA (1996); OES (1997); Comerio et al. (1996)
Notes Dollar amounts rounded.
Programs not mutually exclusive, some recipients received aid from more than one program.

emergency certificates quickly, bypassing many thousands of Los Angeles applicants on a multi-year waiting list for regular (non-disaster) certificates.[3] With Section 8 certificates, disaster victims could rent any HUD approved apartment, and the government paid the difference between 30 per cent of the household's gross income and the unit's rent. The certificate program gave some 22,000 low income victims an opportunity to obtain subsidized housing for eighteen months. More than one year after the disaster approximately 13,000 recipients of the certificates had acquired leases in HUD approved apartments in LA and Ventura counties.

HUD also offered assistance through Community Development Block Grants (CDBG) and Affordable Housing Program (HOME) grants. CDBGs, which are given only to city and county governments, were an important monetary resource for those municipalities seeking funds to develop reconstruction and development programs (Chapter 6). HOME grants support construction of multi-family residences and can provide mortgage assistance for lower income households. While HUD programs have been used after other US disasters, none is specifically designed for disaster relief. HUD made $605 million in grants available to Ventura and LA counties (Galster et al. 1995), including a $200 million supplemental award to fund repairs of apartment buildings in the designated ghost town neighborhoods in the San Fernando Valley (FEMA 1995).

Various other programs were offered by federal and state agencies ranging from mental health counseling to job training and unemployment assistance. FEMA issued grants totaling $35 million to support crisis counseling and mental health programs in Ventura and Los Angeles counties. The Emergency Mobile Home Minimal Repair Program, was developed by the state and funded jointly with FEMA to offer special assistance to mobile home residents. With the California Department of Housing and Community Development (HCD) as the lead agency, a comprehensive plan was developed to reset all damaged mobile homes on their foundations and brace them against future earthquakes, working through a single master contractor. The goal was to cut down on contractor fraud, which was rife after Northridge, speed repairs using approved techniques, and to mitigate against future earthquake losses. More than two-thirds of the 5,000 mobile home owners with damaged units utilized the program in its one year of operation. The program was a departure from other housing programs in that a state agency contracted for the repair work to be done, rather than individual homeowners.

The utilization of the major federal and state assistance programs in the four research communities is summarized in Tables 5.4 and 5.5. MMI average intensity for each site is listed as a general indicator of damage levels. The totals listed for the programs include only the municipalities indicated. Other sections of Ventura County also experienced scattered damage, and the county had a total of 38,000 individual assistance applications (OES 1997). Similarly, while Santa Clarita had the greatest concentration of damage in northern

Table 5.4 SBA loans by municipality

Municipality	MMI	SBA real property	No. of applicants	Average loan	SBA personal property	No. of applicants	Average loan
Fillmore	VII	$1,540,300	73	$21,100	$184,100	61	$3,018
Piru	VIII	$190,100	5	$38,020	$6,900	2	$3,450
Simi Valley	VIII	$53,613,239	2,340	$22,912	$8,104,190	1,823	$4,446
Moorpark area	VII	$6,335,200	276	$22,954	$592,050	162	$3,655
Ventura County total		$61,678,839	2,694	$22,895	$8,887,240	2,048	$4,339
Newhall area	VIII	$12,675,201	595	$21,303	$2,305,700	475	$4,854
Santa Clarita	VIII	$33,561,058	1,870	$17,947	$6,600,901	1,325	$4,982
North LA County total		$46,236,259	2,465	$18,757	$8,906,601	1,800	$4,948

Notes

SBA real property are loans to repair structures.

SBA personal property are loans to replace contents.

Data for Piru includes adjacent rural areas.

Table 5.5 Individual assistance by municipality

Municipality	MMI	IFG	No. of applicants	Average award	FEMA	No. of applicants	Average grant	MHRP	No. of applicants	Average grant
Fillmore	VII	$941,757	1,732	$544	$2,746,713	1,732	$1,586	$1,971,217	714	$2,761
Piru	VIII	$117,132	177	$662	$233,431	177	$1,319	$138,342	57	$2,427
Simi Valley	VIII	$6,557,544	22,223	$295	$48,237,522	22,223	$2,171	$39,213,588	13,998	$2,801
Moorpark area	VII	$3,409,307	4,899	$696	$9,006,112	4,899	$1,838	$7,994,472	3,292	$2,428
Ventura County total		$11,025,740	29,031	$380	$60,223,778	29,031	$2,074	$49,317,619	18,061	$2,731
Newhall area	VIII	$3,098,523	9,235	$336	$16,737,947	9,235	$1,812	$13,543,774	5,060	$2,677
Santa Clarita	VIII	$6,730,204	23,751	$283	$48,611,036	23,751	$2,047	$42,049,368	15,356	$2,738
North LA County total		$9,828,727	32,986	$298	$65,348,983	32,986	$1,981	$55,593,142	20,416	$2,723

Notes
FEMA programs include Temporary Housing and Mortgage Assistance.
MHRP is FEMA's Minimum Home Repair grant program.
IFG (Individual and Family Grants) is administered by OES.
Data for Piru includes adjacent rural areas.

LA county, there were unincorporated areas around Santa Clarita that also sustained losses. Of the four sites, Santa Clarita and Simi Valley had the highest proportion of their populations apply for individual assistance, 30 per cent and 22 per cent respectively. Fillmore and Piru each had about 14 per cent apply. The larger suburban communities also had higher proportions of IA applicants receive SBA loans: 10 per cent in Simi and 7 per cent in Santa Clarita, in contrast to 4 per cent in Fillmore and 2 per cent in Piru. These differences are associated with the lower median incomes and higher unemployment rates in the two smaller communities. Overall, average federal loan and grant amounts are relatively consistent across sites in both counties, reflecting the standardization of federal awards.

Resource access and the delivery of assistance

The Northridge earthquake gave evidence of the large and sustained demand for institutional relief programs after urban disasters in the First World. These demand levels are significant in view of the context: the earthquake was of moderate intensity, it involved relatively new structures in lower population density areas of the urban region, and it occurred in a state with strict building codes, and high levels of social protection. Had the hypocenter been 25 kilometers to the south near the densely populated urban core and the large lower income neighborhoods nearby, the loss levels and assistance needs could have been much higher (OES 1997). The application rates reflect substantial needs and the unprecedented availability of assistance. Northridge marked an effort by FEMA and OES to provide both information and services for an extended period. Most assistance programs accepted applications for at least a year, and several Earthquake Service Centers were kept open for sixteen months.

Over the last decade, federal responses to major disasters have been subject to intense media scrutiny and public criticism. Criticisms commonly focused on delays in federal mobilization, inadequate personnel, inability to serve non-English speakers, the complexity of application procedures (red tape), and delays in receipt of assistance. Having recently borne the brunt of heavy criticism for the 1989 Loma Prieta earthquake and the 1992 Hurricane Andrew disaster (Morrow and Peacock, 1998), FEMA and other federal agencies responded to Northridge quickly, with a large staff of managers and a new-found flexibility in dealing with programmatic restrictions. Federal and state programs were accessible through the DACs and by telephone through FEMA's central teleregistration system, although early demand far exceeded processing capacity, and it took more than a week to get sufficient staff, facilities, and telephone lines to handle the load. Although DACs were heavily used in the first two weeks, with long lines of applicants and overworked staff, teleregistration accounted for 72 per cent of all applications. FEMA/OES opened twenty-one DACs in the two counties, with facilities adjacent to all heavily damaged areas (see Figure 3.5). In addition to application processing, the DACs also provided mental health counseling for victims and referrals to local programs and services.

As DACs were phased out, beginning in week four (mid-February 1994), their services were replaced by eleven Earthquake Service Centers, which were opened over a two-month period. The ESCs offered the same 'one stop' application functions, but represented a new effort by federal and state governments to have a long-term presence in the affected areas. The model for these long-term community centers came from recovery centers OES developed after major firestorms damaged wealthy foothill communities in the Los Angeles region in 1993. The ESCs represented an effort to create centers for community recovery and preparedness activities under the leadership of the center managers, although none evolved into the kinds of long-term community resource centers that developed after the 1993 firestorms. Nevertheless, they provided an extended FEMA and OES presence, offering assistance into the second year of the disaster. As the ESCs were closed, beginning at the first anniversary, their services were taken up in Los Angeles (city) by three Earthquake Recovery Centers. INFO LINE, a non-profit LA County social service organization, also offered a telephone-based information and referral service that continued to assist disaster victims with their recovery needs (Wallrich 1996).

A notable aspect of the Northridge earthquake was the sustained rate at which Individual Assistance applications were received. The initial three-month application deadline was quickly extended on several occasions due to protracted demand, and ultimately terminated, for most programs, on the first anniversary of the earthquake (actually 20 January 1995). Even after that point, FEMA continued to accept applications in cases with 'extenuating circumstances' well into the second year of the disaster. While applications peaked with nearly 300,000 by the end of February 1994, they continued at a rate of 5,000–10,000 per month until the termination of the application period (OES 1997). Thus, while the first two months saw nearly half of all applications (those in areas of the highest shaking intensity tended to apply earlier [OES 1997]), another 300,000 applications were received over the next ten plus months.

As discussed in Chapter 3, a variety of circumstances drove people to seek assistance relatively late in the disaster. For example, homeowners or inspectors occasionally underestimated the cost of repairs, and later learnt from contractors that the costs exceeded their financial resources. In response, they sought out additional loans or grants to cover the costs. Likewise, insurance settlements were sometimes inadequate to cover actual damages, forcing households to seek additional loans or grants. In addition, the 10 or 15 per cent deductible on insurance policies[4] made obtaining loans a necessity for some policy holders to meet the deductible (Comerio et al. 1996). Persons living in condominiums and other housing complexes with homeowners'

141

associations were often confronted with large increases in association fees to cover association property losses, forcing many to seek SBA loans. Any of these circumstances could push people into the individual assistance system months after the earthquake occurred. In addition, there were those who either did not know about the assistance or thought they were ineligible and thus entered the system late in the application period. Overall, the pattern indicates that even moderate urban earthquakes can create both a large initial demand and protracted lower-level demands on programs, and that demand is not entirely predictable from data on ground shaking intensities or damage estimates.

While the nature of earthquake damage and the expenses involved may account for some of the extended demand on relief, the fact that OES and FEMA committed to long-term assistance availability was the enabling factor. The extended availability is traceable to the political claims made virtually on the first day of the disaster by top federal officials. Federal willingness to incur significant expenses by extending program deadlines, encouraging assistance applications, and keeping ESCs operational under-scores the political nature of relief after Northridge. To not have done so in light of early public commitments for long-term support would have been politically costly to the Clinton administration.

DISASTER AT THE CENTER:
THE CITY OF LOS ANGELES

Los Angeles is the largest city in the US (based on area), and much of the epicentral region of MMI VII to IX ground shaking was within its city limits. Thus, it is not surprising that, in comparison to other municipalities, Los Angeles reported the most damage and had the largest number of residents apply for assistance. Housing, including rehousing victims and reconstructing damaged stock, was the dominant issue for Los Angeles and most other municipalities with damage (unless otherwise specified, discussion in this section refers to the City of Los Angeles). While there were losses to commercial structures throughout the region, these were relatively minor compared to housing losses and most earthquake insurance monies went to residential claims, not businesses (Roth and Van 1997). Some 330,000 residential units in Los Angeles were found to have some damage (Los Angeles Housing Department - LAHD 1995), and of the 56,500 residential units (apartments, condominiums, and single-family houses) that were red or yellow tagged, 91 per cent and 77 per cent, respectively, were within the city limits of Los Angeles (see Table 5.6).

In the city at least 19,300 residential units were damaged to the point that residents had to vacate. Of those vacated units, fewer than 2,000 were single-family dwellings. By far the largest population displacement was among apartment dwellers, most of whom were concentrated in the San Fernando Valley. The City of Santa Monica (pop. 89,000), with 70 per cent of its residential stock in the form of apartment and condominiums, also experienced significant damage to multi-family housing. It had more than 300 buildings red and yellow tagged, many of them older masonry structures, displacing 3,500 residents (Nigg 1997). Unlike the San Fernando Valley, Santa Monica had virtually no surplus housing available creating significant rehousing obstacles for displaced residents who did not want to leave its rent controlled housing and beach front ambience.

While the numerical scale of housing damage in Los Angeles may create an impression of a general housing crisis, none was engendered for at least two reasons. First, the vacated residential units represent approximately 1.5 per cent of Los Angeles' 1.3 million housing units, and less than 1 per cent of the county's three million units (OES 1995). In contrast, the 200 plus vacated units in Fillmore (6 per cent of the housing stock) constitute a proportionately far greater loss to the community, particularly given the lack of alternative housing there. Second, for Los Angeles the impact of housing loss and residential displacement was largely cushioned by a high vacancy rate in apartments, an unanticipated 'benefit' of the persistent economic recession that has gripped the region for the better part of a decade. At the time of the earthquake there were 54,000 vacant housing units in Los Angeles, allowing many of the displaced to find alternative housing near their former homes (Comerio 1996).

Damage to the housing stock was unevenly distributed across the region, with scattered but localized concentrations of loss. Approximately half of all red and yellow tagged structures were within 20 kilometers of the epicenter, and 90 per cent were within 30 kilometers (OES 1995). Losses were proportionately distributed across all ages of housing, including those structures built after the 1988 enhancements in the Uniform Building Code. Most of the housing exposed to significant ground shaking was built after 1950, reflecting the concentric outward push of suburban development in the region over the last four decades (OES 1997). In general, damage was heterogeneous and unpredictable, varying due to highly localized interactions of geological and topographic features, surface conditions, and building characteristics (cf. Hewitt 1997: 220). The prominent feature of housing loss was that seven times as many apartment units (ca. 49,000) as single-family homes (ca. 7,000) were red or yellow tagged in LA County (Comerio 1996). While far more renters than owners were affected by the earthquake, homeowners received most of the $9 billion plus in private insurance payments, all of the nearly $1 billion in FEMA Minimum Home Repair Grants, and more than $2.5 billion in SBA loans (OES 1997).

While there was significant damage to housing in dollar terms, little of it was life threatening. In the 1995 Kobe earthquake most of the 6,400 deaths resulted from the collapse of multi-story wood frame homes (IFRCRCS

1996). In the Japanese calamity, nearly 100,000 houses collapsed and 7,000 were destroyed in fires (Hiroi 1997; Yamazaki *et al.* 1997). In contrast, only 33 deaths were caused by building collapses in Northridge, half that number in one apartment building very near the surface point above the hypocenter (Shoaf 1997). Fewer than 200 homes were destroyed by fire, most of those being mobile homes. The majority of residential structures in California, and most structures damaged in Northridge, are wood frame construction (OES 1995). However, unlike the Japanese situation, there is little evidence that wood frame structures there are unduly susceptible to collapse in moderate ground shaking. From a life safety perspective, Los Angeles' residential building codes were adequate for an event of this magnitude, although repairing non-structural damage to homes is proving to be exceptionally costly (Comerio 1997). It has been suggested that many building failures in Northridge had more to do with faulty construction, shoddy building techniques, and 'cost-cutting' by building contractors, rather than from building code inadequacies (SSC 1995).

Housing: surpluses and shortages

A key issue for Los Angeles was the distribution of housing losses and the localized housing crises produced by that distribution. Of greatest concern was the highly concentrated abandonment of apartment and condominium complexes; in seventeen neighborhoods scattered across the city, 60 per cent of the rental housing was lost in the earthquake (LAHD 1995). These abandonments raised concerns about neighborhoods becoming 'ghost towns', areas largely vacated by residents as a result of the earthquake damage. The six most heavily damaged ghost towns, all with at least 80 per cent of the rental stock destroyed, were in the San Fernando Valley. Large complexes were also abandoned in Hollywood and the primarily low income Latino and African-American South Central district, south of the central city. The 'ghost town' phenomenon and the special programs developed for them are discussed in the following chapter.

Federal, state, and city officials moved quickly to assist displaced persons after the earthquake. According to Red Cross officials, a significant proportion of the more than 20,000 shelter residents were staying in ARC shelters because they were afraid to return to largely undamaged (green tagged) apartments. This was a reaction first documented after the Whittier Narrows (LA County) earthquake in 1987 when mostly Latino victims, mindful of the catastrophic Mexico City earthquake of two years earlier, refused to return to green tagged apartments until satisfied that no aftershocks would occur (e.g. Bolton *et al.* 1992). To prevent a repeat of the Whittier Narrows problem, FEMA, City of Los Angeles building safety officials, mental health counselors, and Spanish-language translators formed Reassurance Teams that went to shelters and patrolled neighborhoods, to encourage thousands of fearful residents back into their safety inspected apartments. For those householders whose residences were damaged and not habitable, there was an array of assistance programs and services offered through the application centers.

For temporary housing, both FEMA's rental assistance programs and HUD's Section 8 certificates offered support for replacement housing. In addition, using HUD funds, the LA Housing Department (LAHD) offered $500 Emergency Rehousing Grants for displaced households not eligible for Section 8 Certificates. A total of 3,809 grants were issued for $1.9 million, assisting displaced renters to quickly move in to new housing (LAHD 1995). While $500 is less than the average rent on a one-bedroom apartment in the city, that money in combination with FEMA assistance helped moderate income households afford the usual requirement of first and last month's rent plus a damage deposit before moving in to an apartment. The LAHD grant program was one of a number developed by the department as it reorganized itself into a disaster response and recovery organization (Chapter 6). Rental housing for middle income victims was readily available throughout the Los Angeles area insuring that no significant rehousing problems were encountered for those with adequate resources – including most who received FEMA housing assistance. HUD's Section 8 program provided a resource designed to assist qualified low income households with their post-earthquake housing needs, adding a second layer of social protection to FEMA's broader but shorter-term housing support programs.

While there was a surplus of rental housing in Los Angeles, there was not a surplus of rental housing affordable to low income households (i.e. those living at less than 80 per cent of the $36,000 median household income in Los Angeles). The lack of affordable housing goes well beyond the effects of the earthquake and connects with the economic restructuring that has occurred in Southern California combined with large federal cutbacks in support of housing (Wolch 1996). HUD's Section 8 program, while not a 'normal' part of federal disaster response, provided important assistance to households vulnerable to the loss of affordable housing. HUD specifically targeted very low income households for Northridge Section 8 support, in this case those earning 50 per cent or less of the city's median household income. In one survey of HUD Emergency Section 8 recipients (reported in Comerio 1996: 89), 62 per cent earned less than $10,000 annually, well below the poverty threshold for a three-person household. HUD and FEMA housing programs were not mutually exclusive and Section 8 recipients often received temporary housing rental assistance from FEMA until they were able to find approved housing for HUD certificates.[5] The major problem for low income HUD recipients was finding housing they could afford under certificate provisions and finding landlords who would accept the Section 8 certificates for the eighteen-month period they were initially in effect for.

As Wolch (1996) has documented, while the number of low income households in Los Angeles is increasing, the rental housing stock affordable to such households is shrinking, resulting in a growing mismatch between housing stock and demographic composition of the low income population.[6] These housing problems are exacerbated by the fact that many in the low income population are immigrants and/or ethnic minorities and have relatively large households. The HUD Section 8 Certificates for disaster victims provided a way for at least a small fraction of LA's low income population to acquire rental housing they otherwise would not be able to afford. Of the 22,000 certificates distributed, 13,000 recipients were able to acquire leases on HUD approved housing. One effect of the HUD subsidies was that recipients were able to reduce (though not eliminate) overcrowding by renting larger more expensive apartments, generally with two bedrooms rather than one (Comerio 1996). This, in turn, pushed up local rental rates and created a shortage of larger apartments. (The escalation of rental rates is limited by the LA Rent Stabilization Ordinance, which controls rents on all apartments built prior to 1978 and thus helps maintain some availability of affordable housing.)

In order to place Section 8 recipients in approved apartments, HUD, LAHD and the Housing Authority of the City of Los Angeles contracted with a number of local NGOs and CBOs with expertise in Section 8 housing. Under the 'Mobility' program, Section 8 recipients received housing location assistance from the LAHD and a network of non-profit organizations. Case-workers assisted low income households with a variety of support services including transportation to the DACs, identification of appropriate housing in their neighborhoods, and assistance in negotiating leases (LAHD 1995). The Los Angeles Family Housing Corporation, a CBO that assists the home-less, and others including Catholic Charities and the St Joseph Center, received HUD funding to provide services to Section 8 households. While HUD's programs reached far fewer people than FEMA's 119,000 temporary housing grants, the HUD program offered long-term rent subsidies and provided social support services that enabled very low income victims to acquire new housing. The contracting of case management services to area organizations with local knowledge of housing needs increased the effec-tiveness of HUD's program and the speed at which people were placed in apartments.

A longer-term concern, expressed by both agency officials and community workers, was what the Section 8 households would do when the support period ended (30 June 1995). At issue was that thousands of the subsidized households would be homeless again at the termination of support, unable to afford current rents on their apartments. Anticipating this problem, HUD developed a 'Mobility Plus' program, again contracting with CBOs to begin moving Section 8 households into permanent housing before support expired. Both the Mobility and Mobility Plus programs worked collaboratively with the LAHD, which assembled a database of all available rental units in the city. The housing department also offered a free Housing Information Line, a telephone-based service to assist victims in locating appropriate rental housing (LAHD 1995). The Mobility Plus program, which helped more than 6,000 low income households find non-subsidized housing, became essentially moot as HUD twice extended the emergency Section 8 deadlines for six months (on 31 December 1995 and 30 June 1996). Then HUD converted the approximately 7,000 remaining emergency Section 8 certificates to regular (permanent) certificates, mooting the efforts of the Mobility Plus relocation program. Although HUD's Section 8 program provided a permanent[7] solution to the housing problems of a small group of vulnerable households, it also raised equity issues as the 7,000 households were able to bypass the lengthy waiting lists for regular Section 8 subsidies.

A second Los Angeles housing issue concerned condominium complexes and homeowners' associations (HOAs),[8] and attendant problems of funding and reconstruction. While some condominium complexes were parts of officially designated 'ghost towns' (Chapter 6) dozens of other small and large complexes were spread across the damaged region. More than 12,500 condominium owners in Los Angeles and 780 in Ventura county applied for FEMA individual assistance after the earthquake – approximately 2 per cent of the total IA applications in each county (OES 1997). Many condominium owners had moderate incomes, were elderly, and had no earthquake insurance (LAHD 1995). The complex ownership pattern, in which households own individual units while HOAs owned the building and the grounds, complicated funding for repair as FEMA programs only apply to individual owners or renters, not associations.

The concern was that condominiums would be abandoned piecemeal by homeowners who could not afford repairs or pay large increases in homeowners' association fees charged to fund repairs. The issue was complicated by uneven recovery, in that some owners made repairs while their neighbors abandoned their properties, leaving banks to foreclose on empty damaged units and lowering the market value of the occupied units in the complex. (California has a law that allows homeowners to default on a mortgage without officially declaring bankruptcy, thus making it easier to 'walk away' from a house or condominium that, for whatever reason, they can no longer afford.) Thus, a number of uninsured condominium complexes in Santa Monica, the San Fernando Valley, Santa Clarita, and Simi Valley experienced problems with financing and reconstructing damaged units. A key factor in abandonment were major increases in HOA fees (to $10,000 or more for the year) as the associations attempted to fund repairs or pay insurance deductibles on common property areas. Owners who had little or no equity and who could not afford the increased fees, found it advantageous to abandon their unit. FEMA's MHRP was of little help as it does not pay for repairs to property held by associations.

To prevent abandonment, the SBA began an active outreach program for HOAs and individual condominium and townhouse residents. More than

2,000 HOAs in the two counties were unable to acquire funding or credit to make repairs to common property without major increases in fees to members. SBA provided detailed information and financing strategies to assist in the reconstruction of the complexes. In the first year, more than 500 HOAs received SBA loans to repair common property and another 16,700 individual owners received SBA loans to repair their units (FEMA 1995). While the SBA loans did not eliminate increases in homeowners' fees they did help keep the costs at manageable levels. However, SBA required that residents would guarantee payment on the loans for unoccupied units, that a majority of owners would remain, and that the entire complex would be repaired (LAHD 1995). Some HOAs and condominium complexes could not qualify for SBA assistance under such conditions. The Los Angeles Housing Department, with HUD funding, developed a loan program for those complexes denied SBA loans. The special features of the condominium loan program included income-sensitive variable interest rates and a six-year delay on first loan payments, making them attractive to lower income owners. Condominium complexes within the 'ghost towns' were eligible for a special $22 million HUD funded loan program offered by the LAHD (Chapter 6).

NGOs, CBOs and unmet needs

In addition to the government-based system of assistance, a network of non-governmental organizations (NGOs) and community-based organizations (CBOs)[9] offered a second level of social protection. After Northridge the non-profit sector became active in assisting vulnerable populations who either were not eligible for FEMA assistance or not adequately provided for. In the US, a network of National Voluntary Organizations Active in Disasters (NVOAD), comprised some twenty-six organizations, including the Red Cross and a number of religious groups, coordinates the delivery of volunteer services in disasters. Two state-level VOADs cover Northern and Southern California and LA County has its own network (see Chapter 6).The ARC as the federally designated lead NGO in official relief operations, coordinates a Disaster Network which works with a range of non-profit organizations to provide volunteer staff and material assistance to mass shelter operations.

More than 100 social service organizations in Los Angeles and Ventura counties extended their normal activities, providing emergency services and short-term assistance to households affected by the disaster. Many of these organizations also became involved in servicing client populations who were being marginalized in the relief process. The inaccessibility of relief for very poor households and homeless people was documented after the Loma Prieta earthquake (e.g. Phillips 1993). While there have been improvements in FEMA's service to culturally diverse populations, already homeless persons and undocumented immigrants were largely excluded from the institutionalized relief system.

NGOs and CBOs across the region assisted clients who failed to receive adequate federal assistance for their needs, and they helped marginalized persons who could not enter the official system. The latter included people homeless before the disaster, who, lacking an address, could not qualify for FEMA disaster assistance, and undocumented immigrants who could not seek federal assistance for fear of deportation. The urban homeless – who lack the basic human need of shelter – are among the persistently vulnerable people whose daily lives resemble a disaster. The illegal status of un-documented immigrants inhibited their use of assistance as they live under constant threat of official discovery and deportation. While the ARC offered non-discriminatory services for all victims, irrespective of residency or housing status, accounts from various CBOs indicated that undocumented immigrants were hesitant to use ARC services because LA shelter staff were uniformed. The NGOS and CBOs that worked with immigrants brought qualified clients into the federal/state assistance system by providing referrals and counseling as to their legal rights and entitlements.

The large population of chronically homeless and marginally housed in Los Angeles normally receives assistance from a number of NGOs including the Los Angeles Family Housing Corporation (LAHFC), the North East Valley Homeless Prevention Program, the St Joseph Center, and Catholic Charities (also active in Ventura County). Because the conditions of the homeless population were often worsened by the earthquake, these charitable organizations provided extended services, including shelter facilities, food, clothing, and medical assistance. Local organizations were also active in helping the homeless become rehoused through intensive case management and the provision of transitional assistance in cooperatively owned housing. It is difficult to estimate the numbers of homeless served by this network, but the LAHFC reported serving an additional 1,500 homeless after the earth-quake (Comerio 1996: 103) The homeless are emblematic of a significant marginalized population, people who have, for various reasons, pronounced difficulties in acquiring and maintaining housing (Wolch 1996). (While both FEMA and HUD offer federal support to homeless shelters, in 1997 the governor of California eliminated state funding for a program that provided homeless shelters in California National Guard facilities during the winter.)

A number of Los Angeles NGOs normally serve particular low income ethnic groups, typically Latino or Asian. Perhaps the largest is Catholic Charities, which operated support services to low income immigrants and residents and provided referrals to the federal disaster assistance system. Both the Los Angeles and Ventura County Catholic Charities facilities provided an array of disaster assistance services for immigrants without legal residence. Other CBOs, including El Centro de Amistad and Meet Each Need with Dignity (MEND), which normally provide health and education services for low income Latinos, also offered shelter referral and other support activities for immigrants fearful of going to the DACs (Comerio 1996).

As many of the community organizations served low income and marginally housed people, their clients were often eligible for Section 8 Certificates and FEMA temporary housing. As a result, these organizations frequently assisted resident victims in accessing the official assistance system. Several became actively involved providing additional assistance to households with continuing unmet needs. In the San Fernando Valley, an Unmet Needs Committee was established with representatives from FEMA, OES, and the Red Cross, along with various participating non-profit organizations including Catholic Charities, Adventist Community Services, San Fernando Valley Interfaith, United Methodist Committee on Relief (UMCOR), and Christian Reform World Relief Committee.

As shortcomings in assistance programs became apparent in the first few months of the federal operation, the Unmet Needs Committee began regular meetings to review cases and match people with alternative resources. Both FEMA and OES sent liaisons to the meetings in order to provide official verification of federal and state assistance households had received or might be eligible for. The 'entry qualification' for unmet needs support was that either victims had been denied IFG, or the money received did not cover their needs. The FEMA representative could determine whether individuals might be eligible for further federal assistance due to errors in their applications. While some unmet needs cases were referred back into the federal system, more commonly cases would be referred to one of the participating CBOs that had available resources in the area of need. While the Unmet Needs Committee did not keep records of the numbers of cases processed, FEMA informants indicated that 100–200 cases were reviewed at each tri-weekly meeting. Thus over the committee's year of operation, 2,000 cases may have been reviewed and referred to local organizations for additional assistance.

Most of the non-profit organizations operating in the region offered relatively specialized services for specific populations (e.g. elders, low income ethnic groups, homeless, the physically disabled), and thus were not able to provide the broad range of services needed to compensate for gaps in the federal system. Although the federal government was generally responsive to the problem of unmet needs, statutory limits on program eligibility and the homeowner orientation of its major programs made the network of non-profits an essential, if underfunded, component of the disaster response system. While some NGOs received contracts or grants from government agencies to perform specific services, much of the funding for the non-profit sector was through private sources, particularly individual donations. Well-established national organizations such as Catholic Charities are able to tap a much larger resource base than more specialized CBOs.

The Los Angeles County Department of Mental Health, with more than $20 million in FEMA funding, offered mental health crisis counseling services in Project Rebound, following the model of services offered after the 1992 riots (a parallel program in Ventura County is described below). These services were delivered across the county for more than a year through subcontracts to thirty-six CBOs and NGOs that had local staffs and an established presence in neighborhoods and communities. Organizations such as El Centro de Amistad worked with low income Latino/a clients in the San Fernando Valley, and a similar counseling service was provided for Korean immigrants through the Korean Youth Community Center (Comerio 1996). Other providers offered direct counseling services in the DACs and ESCs, along with outreach programs through LA county schools. School children were targeted as a method of involving entire families who were experiencing adjustment problems after the disruptions of the disaster. The mental health counselors also offered referral services to help clients obtain necessary financial or material assistance. Project Rebound provided services to more than 160,000 clients in the county through its various providers and operations (FEMA 1995).

Los Angeles County-based INFO LINE was a major provider of free information on available health and social support services in the region. It is a telephone-based information and referral service that normally serves 200,000 per year, and its 65-person staff and operations are funded by private donors and LA county (Comerio 1996). As it maintains a large information database and has a multilingual staff, it rapidly became a key referral service for victims after Northridge. It created a special 'disaster line' that provided information on disaster related services to approximately 15,000 callers in the first two months. In the longer term, it worked with the Unmet Needs Committee and provided comprehensive information on available services, helped coordinate information exchange among CBOs and NGOs, and worked in the Mobility and Mobility Plus programs for Section 8 housing. It has continued its information clearing house function long after the 'official' disaster response ended, assisting marginalized populations find available assistance, while helping prepare the non-profit sector for the next disaster (Wallrich 1996).

DISASTER ON THE MARGINS OF LOS ANGELES

In the peripheral areas – including the Santa Clarita Valley and Ventura County – the problems faced in managing the Northridge disaster were broadly similar to those in more heavily urbanized Los Angeles. Yet there was also a specificity to each area's experiences due to demographic and socio-economic differences, location relative to emergency services, and differences in damage and resource access. The sheer size and scale of Los Angeles and the magnitude of the damage within its boundaries overshadowed the impact in smaller, but equally disrupted, communities. Some published damage maps exclude the Santa Clarita Valley and Ventura County entirely, focusing only on Los Angeles (e.g. *Los Angeles Times* 1994). Similarly some published reports under-represent or exclude damage data from some Ventura County communities altogether (e.g. Comerio 1995). It was not until 19 January that Ventura County was included in the presidential emergency declaration, although it was part of the governor's original declaration. These symbolic 'geographies of exclusion' (Sibley 1995) parallel the media's focus on San Francisco and neglect of Santa Cruz and Watsonville after the 1989 Loma Prieta earthquake. Invoking the comparison, one Ventura County Supervisor said he did not want Ventura to be 'forgotten' like Santa Cruz was after the 1989 disaster (*LAT* 24 January 1994).

Delivery of emergency services in Ventura and northern LA counties was complicated by limited highway infrastructure and damage to major connector freeways. In Ventura County 375,000 homes lost power when the regional generating station was shut down from the main shock, and all four study communities experienced interruptions in water and natural gas as a result of ruptures in the distribution systems. Ventura County lacks an integrated water distribution system, relying on three different area distributors in addition to more than 100 local suppliers. Eastern Ventura County, including Simi Valley, receives Metropolitan Water District water through a local distributor but since MWD lost two major aqueducts from the north, it had to rely on emergency reserves. Because of damage to purification systems, residents throughout the county and in parts of Los Angeles County were advised to boil drinking water as a purification measure, in some areas for up to two weeks.

Power failures complicated communication and dispatch of emergency services in the early hours of the emergency, but Ventura County's emergency services operations are experienced in dealing with major disasters. At the time of Northridge, Ventura County was in the process of implementing its Standard Emergency Management System (SEMS) to coordinate activities amongst emergency services departments and to facilitate interactions with the federal agencies utilizing the Federal Response Plan. Within forty-five minutes of the initial shock, the county's Emergency Operations Center (EOC) was in operation, coordinating municipal emergency response activities. While the Ventura County Fire Protection District provides county-wide fire protection services, its central dispatch system failed and calls for assistance were diverted to the area fire stations for direct response. The county's EOC also liaised with the DFO in Pasadena, providing it with information on Ventura's activities and needs. Representatives of two Ventura County emergency response agencies reported that the information requested by the DFO was redundant and time-consuming to compile. The problem, according to one manager, was that the DFO's agencies were not sharing information, thus the county repeatedly had to provide the same information to various agencies. One county emergency manager who had frequent contacts with the

Table 5.6 Building damage and safety data by municipality

Selected municipalities	Total buildings inspected	Residential				Commercial				Other buildings			
		Red	Yellow	Green	Other	Red	Yellow	Green	Other	Red	Yellow	Green	Other
Los Angeles (City)	85,997	1,604	7,715	70,035	272	445	1,105	4,738	8	9	21	41	4
Santa Clarita	4,939	83	184	4,134	137	38	37	272	7	3	5	37	2
Fillmore	508	144	228	5	2	11	17	0	0	45	56	0	0
Simi Valley	6,922	25	299	1,271	2,513	32	150	623	1,873	3	17	98	18
Piru	65	16	33	0	0	16	0	0	0	0	0	0	0
Other Ventura sites	1,571	90	190	904	16	8	6	27	2	3	31	210	103
Total Ventura County	9,066	275	750	2,180	2,531	67	173	650	1,875	51	104	308	121
Total LA County	105,019	1,878	8,495	78,983	2,687	588	1,267	5,545	246	193	743	2,721	1,675

Source: OES (1995)

Notes

Total Los Angeles County includes 37 cities not listed here.

Residential refers to numbers of structures including apartment buildings and single family dwellings

Pasadena office said 'the DFO was chaotic' initially, and others expressed frustration at the lack of coordination among agencies working there.

Damage to Ventura County is estimated at approximately $1 billion, with the greatest losses concentrated in Fillmore/Piru and Simi Valley (Table 5.6). Damage was also reported in low density residential communities in the southern part of the county, including Thousand Oaks and Moorpark. While there were no confirmed earthquake related deaths in Ventura County, there were 798 reported injuries, more than half of these in the Simi Valley area (420 were treated in the Simi Valley hospital and seventy required hospitalization).

Local contexts and city actions

Three years later, the events of the early morning earthquake remained clear in people's memories. Informants emphasized the sounds and the darkness that accentuated their fears. Many told us that the violent ground shaking convinced them that this earthquake was 'the big one'. A Simi Valley man said:

> At first, it was like somebody in a waterbed giving a gentle shove; it felt just like a gentle wave. I felt, maybe about three waves, and I sat up, thinking, 'Oh, neat! An earthquake! Let's sit up and ride this one out.' And then the next thing . . . It felt like a bomb going off. The shaking didn't increase. It was a solid hit. Then, after the hit, the violent shaking set in. What I remember is sitting right here where I am now, and the lights were out. I mean, total pitch black. All I could hear was glass, exploding glass. I first thought the windows were blowing out. And I thought, 'If the house is moving that much, to pop the windows out of the frame and crack them, I'm getting out of here.' I made a beeline for the front door. The house was pitch black. The freezer door in the kitchen had swung open, I ran into it. The furnace door had swung open, I hit it. The closet door was open, I hit it. Then coming up to the front door, I'm thinking, 'I made it to the front door. I'm safe.' I reached up and pulled the door. It opened, probably about a foot, and stopped. And I'm thinking, 'Oh, no. The house is twisted so bad, it won't open anymore. I'm really trapped.'

Families fled from their houses, carrying children and pets with them. They sought out neighbors and clustered in the streets, trying to comfort each other and locate friends and family members. Residents throughout the impact zone went into open areas away from buildings, awaiting the first light of day to view the calamity.

Fillmore

Two weeks before the earthquake, Fillmore had held a disaster preparedness meeting with city departments involved in emergency management. The city had been preparing for the possibility of a flood, given its location downstream from three dams and its proximity to the Santa Clara River. City leaders noted that practice drills and planning sessions enabled the emergency management team to mobilize quickly and efficiently after the earthquake. The small Emergency Operations Center in Fillmore was activated at 5:00 a.m., staffed by local fire department personnel, law enforcement, and city employees responsible for buildings and infrastructure. The county ARC established a triage center and opened two emergency shelters near the EOC within hours of the main shock. More than 350 persons stayed in the shelters in local schools on the first night of the disaster. As building inspectors began tagging unsafe buildings, an additional 250 who vacated unsafe structures went to the shelters. A DAC was established in Fillmore on 20 January and the Red Cross opened a Disaster Service Center in Fillmore on 22 January.

Within the first week, the mostly Latino refugee population at the Red Cross shelters began a steady decline as they started returning to damaged homes or moving in with friends or family. Several hundred persons were still sleeping in tents and cars near their damaged homes in spite of the availability of shelter facilities. A local Red Cross manager indicated that a number of undocumented immigrants were homeless but that they would not go to the ARC shelters out of concerns over government detection. As inclement weather threatened, ARC outreach volunteers went to neighborhoods to encourage people in makeshift shelters to move into the Red Cross facility. The shelter population again increased to 300 per night by the end of the second week as a winter storm struck. Unlike Los Angeles, with its surplus rental housing, Ventura County had a tight housing market with a 1–4 per cent housing vacancy rate by municipality. The fact that many victims were lower income agricultural workers tied to community and work in Fillmore complicated their rehousing, given the absence of alternative housing. Of the 50 persons with reported injuries in Fillmore, 25 were treated at the Santa Paula hospital, a few kilometers west of town. In addition, a mobile medical clinic was dispatched to Fillmore to provide medical services after a clinic (*Clínicas del Camino Real*) that specialized in health care for low income Latinos was seriously damaged and its office building in downtown Fillmore red tagged.

The earthquake caused significant damage across the eastern half of Fillmore and in the downtown business district, damaging its predominantly URM structures, including the Fillmore Hotel, where some 100 migrant workers were housed. Seventy commercial structures in the business district and 510 houses and mobile homes were red or yellow tagged,[10] more than 10 per cent of all housing units in the town. In a nearby mobile home park, 100 units were thrown off their foundations in the main shock. Two months after the earthquake damages were estimated at $50 million, down from the preliminary damage assessment of $250 million. For a small community, with a severely damaged business district and a proportionately large lower income population, paying for recovery and reconstruction became an immediate concern of householders and local leaders alike.

On 18 January the city council met to plan demolition and reconstruction policies. The issue of condemning and razing buildings was complicated by historic preservation advocates who began mounting opposition to demolishing buildings deemed historically significant. Virtually none of the URM buildings in the downtown met current California seismic codes, and the city council had previously refused to make a URM retrofit ordinances mandatory out of cost consideration for owners. Adoption of a downtown redevelopment plan also appeared on the political agenda, given that a Downtown Specific Plan[11] had been approved just prior to the earthquake (City of Fillmore 1994). Political leaders soon moved to adapt the plan to the post-earthquake situation, including retrofit requirements on URM buildings. Thus, within days following the earthquake, long-term business district reconstruction planning and implementation began.

One week after the main shock FEMA Director Witt, HUD Secretary Cisneros and OES Director Andrews visited Fillmore, publicly praising the local emergency response organization and promising long-term federal support for Fillmore and the county. Cisneros, who had already promised major HUD support for Los Angeles, publicly announced that he was authorizing the provision of up to seventy-five mobile homes as temporary housing for displaced households in Fillmore, given the lack of alternative housing in the community. A similar offer was made in a visit to Piru by a high ranking FEMA official two days later. Fillmore began to prepare sites for the HUD trailers, but after two months of uncertainties, local authorities finally determined that the mobile homes would never be sent. Cisneros provided an initial release of $2.8 million in CDBG funds to assist Ventura County with housing and redevelopment. Part of these funds went almost immediately to place a temporary aluminum building structure in a city park to house seventeen businesses displaced from the central business district (Tyler 1997a).

Piru

In contrast to Fillmore's management of the emergency, Piru was not prepared for the disaster and initially overlooked by emergency managers from the 'outside world'. Residents reported significant concerns over the safety of the Piru dam, a few kilometers upstream. It was commonly held that the village had eighteen minutes to escape to higher ground if the dam collapsed. Following the earthquake locals immediately evacuated to the elementary school, the highest spot in town, lacking any information on the status of the dam. Authorities waited three days before informing residents that the Piru dam had not suffered extensive damage. Piru's isolation was a result of both communication interruptions that occurred throughout the county and the severing of Piru's main highway link to Fillmore. The major connecting highway was blocked when a natural gas pipeline ignited and blew a 30-foot crater in the road. For several days residents obtained information by listening to their car radios, since the electricity was off and no other communication media were available apart from some functioning cell phones.

Emergency operations in Piru were handled by its volunteer Neighborhood Council, which established an informal shelter in the local school. However, supplies and resources were extremely limited and external resources became necessary almost immediately. Red Cross, sheriff department officials, and a County Supervisor toured Piru on the day after the temblor but at that point did not carry out any systematic assessment of damage. In the absence of any information to the contrary, county personnel assumed that Piru did not need immediate assistance, and Piru received no significant outside assistance for four days. Hearing of the relief services that converged on Fillmore, Piru residents felt overlooked. As one Piru woman reported:

> They [the ARC and Sheriff's Department] came through and asked if everybody was all right. Well, sure, we were all right. We were all right physically, but we were out of food and water. They didn't even know that we had no telephones, no gas, no electric power . . .

The lack of any operational ties to the county EOC resulted in no communication and delays in preliminary damage assessments. A Neighborhood Council member telephoned the Ventura Red Cross to ask for assistance, but the ARC declined the request since it lacked specific 'official' information on Piru conditions. Once damage information was conveyed to a County Supervisor, she in turn, was able to provide emergency services personnel who took over service delivery from the informal response system that had emerged.

Piru's small central business district of fourteen URM buildings was heavily damaged in the temblor, and only two small food markets remained operational. Piru's only bank was closed permanently as a result of damages, making trips to Fillmore a requirement for any banking transactions. More than 10 per cent of Piru's housing stock, mostly old wood frame homes and mobile homes, was seriously damaged, with 16 red tagged and 33 yellow tagged structures. The estimate of total damage in Piru has been set at $3 million. Apart from the displaced businesses, employment was largely unaffected as most residents worked on area farms or in a local fruit packing facility that remained operational.

As an unincorporated community, Piru lacked any resource base required to initiate earthquake recovery, relying solely on County resources. Given marginal economic situations, Piru's small business owners failed to meet requirements to qualify for SBA loans and owners initially resigned themselves to losing their buildings. However, through the intervention of the County coordinator for Piru's earthquake recovery, a retired Ventura County Administration employee, Piru received a $450,000 federal historic preservation grant and other resources to help rebuild its historic business district (Chapter 6). The coordinator also obtained three trailers within which local businesses could operate during the reconstruction of the downtown business district.

Simi Valley

Simi Valley, located 16 kilometers from the hypocenter, was subjected to somewhat more severe ground shaking (up to MMI VIII) than Fillmore and Piru. The city suffered infrastructural damage to highways, streets, and water storage facilities, and water service and sewage were interrupted for up to four days in some areas. The most serious highway damage was a collapse and closure of the Simi Valley Freeway into the San Fernando Valley. There was structural and nonstructural damage at twelve public school buildings as well as at the city's administrative building and the police department.

As a result of its location, as well as size, demographic make-up, and diversified economy, Simi Valley's response was markedly different from Fillmore and Piru's. Simi's economic base is a combination of high technology firms, banking and insurance centers, low-wage garment manufacturing, and numerous large retail businesses. It is a relatively new community with no URM structures, although its several mobile home parks were extensively damaged. Its proximity to the San Fernando Valley also makes it a convenient 'bedroom community' for commuters into Los Angeles. Connected to LA by a Metrolink railroad line, Simi was not isolated after the earthquake and residents could continue to commute to work in the San Fernando Valley in spite of heavy damage to freeways.

The City of Simi Valley had its own EOC managed by the police department, which handled operations in its jurisdictional area in coordination with the County EOC. The city EOC was mobilized within thirty minutes of the temblor, and the police department activated the earthquake response plan and dispatched units to assess and report damage. Mutual aid was requested through the County EOC for fire, police, and building safety personnel to assist in providing needed emergency services. Damage to the operations center did impair early response activities, although there were no major reported difficulties.

A total of 575 residential units were red or yellow tagged and an additional 622 mobile homes were damaged, taken together amounting to less than

5 per cent of the total housing stock (Figure 5.6). Ground failures combined with flooding from ruptured water mains and collapsed water tanks concentrated residential damage in the eastern end of the city. A total of 175 commercial units, including several large retail structures, were damaged in the main shock. Estimated losses to public structures, homes, and commercial buildings total $418 million, accounting for nearly half the total dollar losses in Ventura County. Estimates of the displaced residential population were as high as 4,000 (out of 100,000), although only 200 used the ARC shelter in the local high school. As in Fillmore, the ARC sent representatives into the neighborhoods with building inspectors and mental health counselors to encourage those camping in yards to come to the emergency shelters. The Red Cross also provided direct food services to some of the 'seniors only' mobile home parks nearby. The city's EOC ceased emergency activities one week after the event, and most participating departments and agencies went back to routine operations.

FEMA/OES opened the first Disaster Application Center in Simi Valley on 20 January, three days after the main shock. This DAC was rapidly overtaxed as FEMA bussed in San Fernando Valley victims to take pressure off heavily used DACs in that area. A second DAC was opened one week later, and both DACs continued in operation until mid-February when they were combined into a single Earthquake Service Center. The Red Cross shelter remained open for three weeks, by which point use had dropped off significantly. According to ARC shelter managers, upwards of one-third of the shelter population in Simi consisted of people homeless at the time of the disaster. Working with the Simi Housing Authority, the Red Cross located temporary housing for displaced residents comparatively quickly. There were more temporary housing options available in Simi Valley than in Fillmore, and Simi Valley victims were more easily placed in temporary housing. Given that the poverty rate in Simi Valley is less than a quarter of Fillmore's (Table 4.1) Simi victims did not face the same economic constraints that the rural areas did. San Fernando Valley residents seeking new housing in Simi Valley placed additional pressure on housing, but authorities reported that housing remained available for middle income households.

Santa Clarita

Santa Clarita was the most heavily damaged city outside of the San Fernando Valley. Like Simi Valley, it is a newer bedroom community with a relatively 'young' housing stock. Santa Clarita was incorporated in the last decade, encompassing several existing communities including Newhall, Saugus, Valencia, and Canyon Country. Located 20 kilometers northeast of the hypo-center, parts of Santa Clarita were subjected to MMI VIII ground shaking (OES 1995). The community was without power for one day and significant damage to water purification and delivery systems cut off potable water to unincorporated areas of the Santa Clarita Valley, in some areas for a week or more. An oil pipeline ruptured spilling 200,000 gallons of oil into the Santa Clara River channel near the city. There was extensive damage to mobile home parks, with twenty units destroyed by gas fires, and 1,700 knocked off their foundations and damaged.

Santa Clarita was also affected by significant highway damage to Interstate 5 (I-5), a major north-side traffic link that daily channels hundreds of thousands of vehicles past the city. Bridges in Gavin Canyon south of Santa Clarita collapsed, resulting in severe traffic congestion within the city as traffic was re-routed on city streets. For automobile commuters Santa Clarita was effectively cut off from the San Fernando Valley and twenty-minute commutes turned into four-hour nightmares on alternate routes. Additional highway and bridge damage further limited traffic flow in the spatially dispersed city. As in Simi Valley, commuting problems were mitigated by the presence of a Metrolink rail line into the San Fernando Valley and downtown Los Angeles.

The city opened its EOC almost immediately after the main shock, although it could not be situated in the heavily damaged City Hall building. Emergency operations were transferred into tents and trailers in an adjacent parking lot. As a result of mutual aid requests, 600 additional fire personnel and 175 additional law enforcement personnel arrived on the second day. The ARC opened four emergency shelters, with one housing elderly residents displaced from damaged homes and mobile homes. By the second night, the shelter population was up to 1,000 persons, and an additional 400 were camped at a city park in Newhall. The first DAC was opened on 20 January in Canyon Country, in eastern Santa Clarita. The ARC opened two Service Centers on 28 January to provide housing and material assistance to victims waiting to receive FEMA grants. The following day FEMA opened its second DAC to help ease crowding. FEMA provided the City of Santa Clarita with $7 million in early disbursements to assist it in covering costs incurred in responding to the emergency. The city provided a bus service for victims to access the DACs in order to reduce traffic congestion on the constricted roadways in the area. Both DACs were combined into a single Earthquake Service Center on 18 February in Valencia.

Santa Clarita's building damage was extensive with 50 single-family homes red tagged and 168 yellow tagged. An additional 87 apartment and condominium units were red tagged and 202 were yellow tagged. Including more than 1,700 mobile homes knocked off mounting jacks, approximately 5 per cent of the housing stock was damaged at least moderately. Thirty-eight commercial structures were red tagged, including several large retail stores. Private home and commercial structures were estimated to have suffered $200 million in damage, while additional damage to public property and buildings, including the City Hall, three Post Office buildings, and three public libraries was in excess of $99 million. Damage to highways, schools, infrastructure, and to unincorporated areas brought the total estimated losses in the Santa Clarita Valley to $428 million.

Traffic, highway, and water supply problems continued to trouble the area into the third week of the disaster. Heavy traffic congestion, including thousands of large trucks daily, produced significant deterioration of city streets as I-5 traffic was re-routed through the city. Water supplies in unincorporated areas outside Santa Clarita were slow to come on line, and some mobile home parks were still without utilities in early March. One outlying mobile home park remained without running water into July, relying on portable showers and toilets provided by FEMA. The Red Cross provided a food service to some mobile home parks, where residents were camped, refusing to leave their property for ARC facilities. The three remaining Red Cross shelters were consolidated into a single facility on 6 February. More than 150 were still in the shelter at that time, and they were placed in temporary housing over the following two weeks. As with other areas, the shortage of affordable housing created difficulties for displaced lower income households.

City leaders initiated recovery planning by the second week of the emergency. To assist with initial restoration expenses, the city received a $770,000 early disbursement from HUD followed shortly by $7 million in advance funds from FEMA. As the city staff was familiar with FEMA application procedures from a 1993 flood disaster, they were able to move quickly to acquire additional federal funds to begin recovery activities (Nigg 1997). Being in Los Angeles County also provided Santa Clarita with political leverage in acquiring federal assistance through the efforts of the LA County Board of Supervisors. Drawing on both available financial resources and a skilled city staff, Santa Clarita developed a range of city assistance programs to facilitate earthquake recovery for households. Due to its success in acquiring federal resources, it was also able to move forward on repair of public buildings, including a retrofitted City Hall, repaired at a cost of $4.6 million (City of Santa Clarita 1994a, 1994b).

One problem area that city officials had to confront was identifying and contacting absentee owners apartment complexes in the old Newhall district, an area populated by lower income Latino and elderly renters. Analogous to problems surrounding the ghost town neighborhoods of Los Angeles, the owners of damaged apartments lacked resources or access to credit to repair them and they were at risk of abandonment. Unlike Los Angeles, which maintains an elaborate database of structures and building owners and thus was able quickly to identify owners and arrange financing, Santa Clarita officials had considerable difficulty in locating building owners in Newhall. This led to delays, in some cases, of a year or more, before rehabilitation of buildings was begun. According to building officials, there were numerous cases of lower income renters who continued to live in apartments, although the buildings had been yellow tagged by city inspectors as unsafe for permanent occupancy.

In summary, all four communities moved quickly to respond to the disaster, but their capacity to respond reflected the financial and organizational resources available to them. Despite the region's high per capita income, there are pronounced discrepancies in resources available to households and local governments. Wealthier communities, including Santa Clarita and Simi Valley, could initiate rehabilitation programs not available to the smaller rural communities with their limited resource base.

Local perspectives on assistance

For earthquake victims interviewed in this research, recovery often comprised a long, complex, and occasionally frustrating process. FEMA specifically attempted to service culturally diverse and economically marginal populations, responding to criticisms after Loma Prieta and Hurricane Andrew. Yet, as newspaper chronologies revealed, stories of 'frustrated' victims complaining of complex forms, denial of loans, and inadequate assistance were common features of local media coverage in the first six months of the earthquake (e.g. *Enterprise* 18 June 1994). Among study participants it was clear that those unfamiliar with formal bureaucratic procedures and those with fewest alternative financial resources were more likely to find the process difficult and lacking in satisfactory outcomes. In what follows, we consider examples of lower income homeowners' experiences as they attempted to secure resources through conventional FEMA programs. As one respondent described the application process: 'They were nice to you, but they'd sure push you in circles.'

In spite of the teleregistration system in place, DACs in larger communities (including Simi Valley, Santa Clarita, and most of the San Fernando Valley) were deluged with people waiting in line for assistance during their first week of operation. Out of the 1,500 waiting to use the Simi Valley DAC on its opening day, only fifty-five applicants were processed, with the remainder having to return later that week. Crowding was relieved as FEMA opened new DACs across the region into the second week. Transportation assistance to the DACs was arranged by local social service agencies which helped those households that lacked easy mobility in rural Ventura County. As most households needed to make two or more trips to the DACs to complete all forms and provide necessary documentation, convenient transportation was a major advantage.

In contrast to the heavy demand in the larger communities, service providers noted an initial reluctance among the majority of Latino residents of Fillmore and Piru to apply for any federal assistance. Local agencies reported having to convince Latino residents and legal immigrants to apply for assistance. Rather than going to the DAC, victims obtained assistance from the local Catholic church or, in some instances, the Red Cross shelter, accepting blankets, water, and food, but initially avoiding the 'official' center.

As a Ventura mental health counselor remarked, 'A lot of Latinos around here think the federal government can just load them up in box cars and ship them back to Mexico, no matter how long they've lived here.' Others used terms such as 'fear' and 'distrust' to characterize prevailing attitudes among lower income Latinos toward the federal presence. Over the weeks following the main shock, outreach and information efforts by FEMA, the Red Cross, and local agencies did bring legal residents into the system. However there is no evidence that undocumented immigrants used any federal programs at the research sites, whatever the concerns of conservative ideologues.

Legislative action and public discourse directed at undocumented immigrants in California exacerbated fears of immigrant groups and prejudices of dominant groups toward them. The prevailing anti-immigrant political culture appeared to have its greatest effects in the largely Latino rural areas of Ventura County, with its mix of long-time Mexican-American citizens, legal Latino immigrants, and undocumented migrant workers. The denial of federal assistance to undocumented immigrants, legally mandated as part of the congressional appropriation, was a component of the politics of fear for immigrant earthquake victims. As discussed in Chapter 3, the federal actions were bolstered by California Proposition 187, a controversial measure that passed easily in the November 1994 election. This attempt to deny foreign nationals social services became immediately tied up in the courts as its constitutionality was challenged.

In the small communities of Ventura County, damage to existing housing and the lack of affordable alternative housing created significant barriers to rehousing those with limited economic resources. Informants involved in provision of affordable housing reported that the problem of relocating several hundred low income earthquake victims led, in some instances to red tagged structures being given yellow tags by local inspectors due to the lack of alternative housing in the area. The problem occasionally extended beyond the rural sites. One Simi Valley resident described living in a red tagged structure:

> When the inspector came and saw the house, he said that the house was not safe to live in [red tagged]. But from the staircase, here, this way to the back of the house, was okay. Fortunately, most of our living quarters are in the back. But any time we had to leave the house we went out the front. In order to come in and out of the house, you have to go through the front part of the house. And every time we had an aftershock, everybody would run out of the house that way.

In rural Fillmore and Piru, building assessment was complicated by the fact that damaged houses did not meet building codes even before the disaster. County statutes require that once an inspection has been carried out, all buildings on a property may be required to conform to the building codes.

Thus those applying for assistance ran the risk of being required to bring all structures (garages, out-buildings) on their property in to conformity with current codes, even if not damaged in the earthquake, an expensive proposition for low income property owners. A Piru resident seeking FEMA assistance discussed at length his problems dealing with FEMA building inspectors due to his substandard house. The inspector examined the house in January 1994, providing a preliminary assessment and indicating that a more complete inspection would be conducted the following month. In March 1994, the resident went to the Fillmore ESC, only to be informed that the initial inspection records had been lost. As the resident told his story:

> That must have been when they lost all those inspections. Mine was one of the ones they lost anyway. So, around April, I'm standing here . . . and I get a call from the FEMA lady, and she says, 'Can you come in to the office this afternoon. We'd like to talk to you. We've got your [inspection] report, but you're not going to be happy with it.' So, I said, 'fine', and I went over there. And this inspector had put down that he didn't think we were living in the house at the time of the earthquake! . . . And I says, 'Well, I'll show you the proof.' I showed them the tax receipts and my gas bill and my water and light bill. The FEMA guy said, 'I don't know what the inspector saw . . . but there wasn't anybody in the house.' I said, 'No, we weren't living there when the inspector came because it was unlivable.. It had been ruined by the earthquake.' I said, 'This must be a joke. I can bring you 1,200 people from Piru who know that this was my primary residence at the time of the earthquake.' He said, 'That is what the inspector said. This is what you have to do. You have to write up a letter and ask for a reassessment.' And I said, 'Why do I have to ask for a reassessment?'

The reassessment was scheduled for October 1994, but the resident's home was not reinspected until January 1995. By then he had qualified for a small grant from a Fillmore CBO. However, the CBO required that applicants first apply for FEMA and SBA assistance and be denied. Thus the resident could not receive his grant until his application was processed by FEMA and an official determination made of eligibility. At the time of the interview, conducted in more than eighteen months after the main shock destroyed his home, the informant was still waiting for a final ruling by FEMA. His story was by no means unusual among owners of small lower cost homes in Fillmore and Piru.

Fillmore, Piru, and the Newhall section of Santa Clarita, all had a quarter or more of their residents who were monolingual or bilingual Spanish speakers. DACs had bilingual assistance workers available for applicants, but non-native English speakers appeared to require more than translation assistance to work through assistance applications. In the Fillmore and Piru areas, Spanish speakers were frequently agricultural workers, originally from rural Mexico, poorly educated and not skilled at paperwork. They followed the relief workers' translated instructions on their first visit to the DAC, but they often did not understand the need to maintain all receipts, follow up letters, or spend FEMA money only on designated repairs. According to social service agencies working with this population, there were repeated cases where low income householders were reluctant to spend the assistance money received given the uncertainties they had over what the money could be spent on as per FEMA regulations.

An inevitable feature of large-scale and long-term FEMA responses is that there is a rotation of staff during the course of the disaster operation. While respondents in the suburban communities generally accepted this as a feature of the assistance system, those in the smaller communities were more likely to report that the personnel change-overs created difficulties in their application procedures. Situations were commonly reported in which respondents indicated that they were asked to submit either duplicate or new documentation when their cases changed hands with new personnel at the ESCs. As one Fillmore resident reported in a focus group interview:

> We wouldn't hear from them at the [earthquake service] center for awhile, so we would call to talk to Harry. And they would say, 'Harry? Now Joe has your case.' And Joe would say, 'I need you to send me "x", "y", or "z".' And we would say, 'We already did that for Harry.' And Joe would say, 'Well, now you need to do that for me.' I tell you what I'm finding out in this disaster . . . is that they change managers like you take a shower every day. Every day somebody else comes in, and there's a new command. The only people who are left outside are the outreach workers, and they change every three or four months. But, as far as the manager goes, they come and go like nothing . . . When I finally got in, the guy told me, 'I wasn't going to wait for you come in. Today's my last day, and I'm gone.' So, I come back tomorrow, and I say 'Joe told me I was going to have my inspection.' And they tell me, 'No, no, we're not doing it that way, and now you have to do it this way . . . '

While such instances occurred in a minority of cases, they point to the occasional gulf between bureaucratic 'culture' and the face-to-face informality that rural respondents appeared to prefer. For many residents, the recovery process required a knowledge of bureaucracy and skills in completing paperwork that they did not have. Despite the creation of a large institutional framework of support, pockets of marginal households continued to have problems in obtaining needed financial assistance.

Issues in assistance

It is, of course, inevitable that in processing more than 600,000 applications problems will occur. The stories of the informants in the smaller communities suggest that their difficulties relate, in part, to the fact that they did not necessarily 'fit' the types of households that most FEMA programs were designed for. US researchers have noted that federal disaster programs are designed largely for middle-class homeowners with a continuous record of employment and residency and work less well for those that do not fit this pattern (Bolin 1994a; Comerio *et al.* 1996). Among study participants, those from economically marginal households experienced specific problems that went beyond the occasional lost form or delay in reimbursement that were relatively routine. These appeared to involve difficulties in filing SBA loan applications, problems in getting information and following through on their FEMA applications, and for homeowners, there were difficulties with inspections. These lower income, mostly Latino households also lacked the personal resources or credit to fund their own repairs leaving them in an increasing precarious situation in the absence of substantial federal assistance. The barriers these marginal households encountered in accessing resources marked their lack of resilience in the face of sudden changes in living circumstances. As in Los Angeles, there were a number of NGOs and CBOs that developed programs and offered services to assist households that were not served adequately by federal programs. CBOs and NGOs provided the human and fiscal resources that augmented social protection available to vulnerable households.

While most of the assistance difficulties identified here involved lower income households in the four communities, there were additional issues that pertained to middle and upper middle income homeowners, that delayed assistance and recovery. These primarily involved those homeowners who could neither qualify for SBA or commercial loans to repair damaged homes and have either been unable to repair homes or went into foreclosure. As discussed below, these cases of situational vulnerability depend on particular combinations of events not necessarily related to structural conditions in society. One area of more than a dozen homes in Simi remained unrepaired two years after the disaster due to protracted ground settling that continued to break up foundations and houses. To stabilize the ground took an expensive compacting process that raised the costs of repairs beyond what some owners could afford. In addition, newly drawn flood plain boundaries along Arroyo Simi, placed dozens of Simi homes within the flood plain. Thus houses with at least 50 per cent damage are required by FEMA to be elevated above the flood plain, adding even more to the cost of repair. Thus in the two years following the disaster none of the owners in the area was able to start reconstruction, nor could they sell the homes except at a major loss. Two of the homes had been foreclosed by banks during the course of the research.

In Santa Clarita, economically stressed middle-class homeowners had to cease reconstruction on fifty red or yellow tagged homes due to funding shortfalls, and thus two years after the main shock these homes remained partially repaired and uninhabitable. Another twenty heavily damaged homes were razed and never rebuilt as the owners could not acquire the necessary loans for reconstruction. While more than 70 per cent of the single-family dwellings in both Simi and Santa Clarita were either repaired or being reconstructed by the second anniversary (January 1996), there was a minority of middle income households that lacked access to adequate resources in the face of very high home costs. In the absence of insurance or SBA loans, the costs of rebuilding homes for middle-class homeowners was well beyond the modest resources available from most CBO and NGO programs, leaving owners with few options beyond walking away from ruined homes.

CBOs, unmet needs, and vulnerability

In general terms, NGOs and CBOs were a critical factor in assisting those lower income households that could not acquire adequate, or any, federal assistance. Both FEMA and OES recognized and cooperated with the non-profit sector as they attempted to 'fill in the gaps' in the mainstream programs. The NGOs and CBOs in the research communities attempted to serve households with continuing unmet needs during their recovery. For our purposes such unmet needs can be taken as markers of vulnerability, as they indicate entitlement failures. Unmet needs after disaster are the result of two usually related phenomena: the depth of existing social inequalities that produce vulnerable households, and inadequacies in institutionalized relief programs.

While federal programs were widely available in the disaster area, they were not necessarily accessible by all who needed assistance. Unmet needs in housing, health care, employment, and the rest are components of vulnerable households' daily lives. A disaster such as Northridge compounds those needs, sometimes deepening the disadvantages that people live with. Assisting marginal households with their unmet needs becomes an important issue in recovery. However, disentangling pre-disaster unmet needs from 'disaster-caused needs' and then only offering assistance for those attributable specifically to the disaster is a hallmark of FEMA assistance programs. Ignoring the needs marginal households had before the disaster, as federal programs do, can inadvertently reinforce the structures of vulnerability. The constraints faced by lower income victims, including substandard, overcrowded housing, marginal employment, cultural marginalization, and language barriers worked to restrict access to or disqualify them from most official relief programs. Yet these same constraints produced needs that went well beyond those that can be attributable to earthquake damage. It was to the needs of marginal households that the non-profit sector responded.

167

Church-based NGOs rapidly mobilized after Northridge and concentrated on identifying critical needs of marginal households in the smaller Ventura County communities. Christian Reform World Relief Committee (CRWRC) members and volunteers from the Mennonite Disaster Service arrived in Ventura County within days of the earthquake. CRWRC conducted a survey of Fillmore and Piru residents to identify victims' recovery needs in light of local conditions. The needs assessment was submitted to the local ministerial association to promote the development of local programs and to offer referral assistance for victims to federal programs. While not formally constituted as an unmet needs committee, CBOs and NGOs began targeting households that were unsuccessful in acquiring adequate federal resources. As with non-profit organizations in Los Angeles County, some Ventura County CBOs and NGOs were able to utilize county CDBG funds to support recovery programs for marginal households. As such, it was the County government's ability to 'work the system' to local advantage that created a resource base for local recovery programs (see Table 5.7).

Emblematic of a community-based unmet needs organization, the Fillmore-based organization 'Rebuild: Hand to Hand', known locally by its Spanish name *Mano a Mano*, was organized and incorporated two months after Northridge. It emerged from a coalition of local churches, businesses, and civic organizations from Fillmore and Piru. Its early effectiveness was facilitated by a staff of well-known local residents, an important factor in small communities in which many were uneasy about the federal presence. The director of Mano a Mano, a Fillmore resident, explained the problem:

> The people [in Fillmore] have lots of financial problems, in addition to problems repairing their homes or businesses. Most are low income, with little savings. With the people I've seen, it's very rare that anyone has earthquake insurance. They weren't financially prepared for recovery costs. They don't qualify for regular loans and are dependent on volunteer services and outside sources. Many still haven't received any [FEMA] money and are concerned about how long the process takes.

Mano a Mano acted as the local coordinator for the activities of several NGOs in the area, including Habitat for Humanity and Mennonite Disaster Services. From its establishment, Mano a Mano focused its efforts on providing assistance to low income homeowners. Its approach relied on using its resources to purchase building materials and coordinating the volunteer labor efforts of the Mennonites in repairing homes. A major funding source for Mano a Mano programs came from a grant received through the city of Fillmore's Redevelopment Agency (Chapter 6). The organization received $575,000 from the Fillmore agency to buy building materials for low income victims who could not acquire adequate federal assistance to repair homes. In its first year of operation, the CBO had organized and funded the repair or reconstruction of forty homes in Fillmore and Piru. By purchasing bulk construction materials and drawing off volunteer labor, the organization was able to 'stretch' its funding considerably to assist low income households. Given the hesitancy of Latino residents in the rural communities to apply for assistance, Mano a Mano staff, through their outreach efforts, also helped local victims apply for and follow up on federal assistance (see Table 5.7).

Habitat for Humanity worked cooperatively with Mano a Mano during the eighteen months of the latter's operation. Unlike Mano a Mano, Habitat has been involved in numerous post-disaster housing and development projects in the US and is thus experienced in acquiring resources from donors. Habitat's programs, while reserved for those with unmet housing needs, was not restricted to just lower income households in Fillmore and Piru. It offered assistance to middle income homeowners as well, contingent on them having unmet needs after exhausting federal resources. According to Habitat organizers in Ventura County, there were numerous cases where building inspectors underestimated actual repair costs, thus leaving some homeowners with inadequate grants or loans to repair homes. While Habitat targeted its volunteer-based repair and reconstruction services toward lower income households (below 60 per cent of median income), it did participate in the repair of nearly two dozen homes of middle income residents in Simi Valley. These latter were characteristic of situationally vulnerable households: middle-class homeowners with restricted incomes and no insurance who could not qualify for SBA or commercial home loans. Habitat also undertook twenty major home repair and rebuilding projects for low income home-owners in Fillmore and Piru. To insure residents' participation, recipients of Habitat assistance were required to donate one hour of volunteer service in return for each $100 spent on building supplies on repair projects. A Habitat informant indicated that the volunteer service arrangement worked more effectively with low income households, referring to a tendency for middle income homeowners to treat Habitat volunteers as private contractors.

Mano a Mano ceased operations at the end of 1995 after failing to obtain additional resources to continue its work. Habitat, on the other hand, successfully obtained more than $1 million in block grants and $2 million in donations to engage in development projects in the area, including the construction of twenty-two new homes for very low income residents of Piru (Chapter 6). While NGOs such as Habitat had the expertise to obtain continuing funding, delivery of their services required a cooperative working arrangement with local organizations such as Mano a Mano. In interviews, Habitat for Humanity staff commented that as an 'outside' NGO they recognized they themselves did not know what community residents needed for recovery, and were dependent on local organizations for their legitimacy. Linkages among CBOs and NGOs became strengthened during the course of recovery through their cooperative rebuilding programs.

169

Table 5.7 CBOs and NGOs in the study communities

Name of organization	Community location	Primary activity	Target population
St Francis Catholic Church	Fillmore	Emergency relief provisions	Displaced households
Christian Reform World Relief Committee	Fillmore and Piru	Damage and needs assessments	Low income homeowners
Catholic Charities	Fillmore, Piru, Simi Valley	Material assistance, counseling, referrals	Lower income, farmworkers, undocumented immigrants
Mano a Mano	Fillmore and Piru	Building permits, materials, grants, volunteer labor	Low income homeowners
INFO LINE	Santa Clarita and northern LA County	Information, unmet needs referrals	All residents
Retired Senior Volunteers Program	Simi Valley	Information, social support application assistance	Elders
OASIS – Older Adult Services Catholic Charities	Simi Valley, Fillmore	Social support, counseling, referrals	Elders
Interface	Simi Valley	Counseling and referrals	Children
Cabrillo Economic Development Corporation	Fillmore and Simi Valley	Affordable housing projects	Low income families, farmworkers
Affordable Communities	Piru	Affordable housing projects	Low income, elders
La Clínicas del Camino Real	Fillmore and Piru	Health services and earthquake counseling	Low income households

Table 5.7 continued

Name of organization	Community location	Primary activity	Target population
Habitat for Humanity	Simi Valley, Fillmore and Piru	House repair and building	Low income homeowners
Caring Link	Santa Clarita	Information, referrals, volunteer coordination	All residents
Mennonite Disaster Service	Fillmore and Piru	House repair and building	Low income homeowners
Commission on Human Concerns	Fillmore and Piru	Disaster preparedness training, mitigation assistance	All residents, very low income households
United Methodist Committee on Relief	Santa Clarita	Application assistance, referrals to FEMA, labor assistance	Low income homeowners, mobile home owners, elders, disabled persons

While the NGO/CBO programs addressed unmet needs of low income homeowners, displaced renters in the county faced different housing constraints. As is typical after US disasters, displaced households sought short-term housing accommodations with family and friends in the four research sites. For renters in Fillmore and Piru, finding new housing in the same community was complicated by the shortage of undamaged affordable rental housing. While HUD distributed approximately 600 Section 8 certificates in the county, neither Fillmore nor Piru had any available apartments that were eligible to take the certificates. In contrast to Los Angeles, relatively few Ventura County households could use the program. Large families and those lacking adequate English-language skills often encountered problems in locating apartments that would accept the certificates. Most relied on assistance from the Ventura County Housing Authority to locate qualified apartments and to negotiate leases with landlords unfamiliar with the Section 8 program. As no Section 8 approved housing was available in Fillmore or Piru, those wishing to use the certificates had to relocate to larger cities including Oxnard 45 kilometers southwest of Fillmore. Low income workers could not afford the travel costs or time to move between Oxnard and their jobs in Fillmore on a daily basis. As a consequence of the lack of suitable housing in rural Ventura County, most Section 8 recipients were not able to use them. The Area Housing Authority of Ventura County estimated that fewer than 150 families from Fillmore and Simi Valley had been able to use their Section 8 housing certificates.

Although not on the same scale as poorer communities, both Simi Valley and Santa Clarita social service agencies reported that there were clusters of marginal households with persistent unmet recovery needs. Because these households represented a small portion of all victims in these largely middle-class communities, they were 'nearly invisible, slipping through the cracks' of the assistance system as one social worker described it. A key effort of local social service agencies involved outreach programs in schools, mobile home parks, at Red Cross shelters, and ESCs, attempting to identify households with ongoing recovery needs. As needy households were identified, NGO staffers would provide information and referrals to other service providers in an attempt to help meet victims' recovery needs. Elders in Simi were identified as encountering particular difficulties both in coping emotionally and with repairing homes or finding new housing. OASIS (Older Adult Services Catholic Charities) and RSVP (Retired Senior Volunteer Program) were both active in Simi, providing elders with social support, transportation, and assistance with applications. Habitat also provided repair assistance for Simi homeowners unable to acquire necessary resources through conventional programs. The City of Simi, through a CDBG-funded grant program, assisted households in the purchase of building materials to be used in conjunction with Habitat volunteer labor.

In Santa Clarita, the United Methodist Committee on Relief (UMCOR), a member of the NVOAD, hired staff to coordinate local volunteer services in the area, working in conjunction with the Unmet Needs Committee of Los Angeles County. The UMCOR staff and volunteers worked with disaster victims referred by ESC staff, the local Red Cross chapter, and area social service agencies. UMCOR activities included providing transportation for elder victims, assisting with relief applications, appeals of settlements, and connecting those with unmet needs to locally available resources. UMCOR also coordinated volunteer labor assistance for low income victims' home repairs, and remained an active information and referral service for the northern area of LA County. As their main clients were a mix of lower income homeowners, elders, and mobile home residents scattered across the area, it took considerable effort to locate and provide services to them. According to UMCOR staff, their elderly, low income, and ethnic minority clients had been 'just barely surviving' before the earthquake and they contended that without their 'personalized' services, the living conditions of these vulnerable households would have deteriorated significantly. A common concern of NGO/CBO staff in the larger communities was a shared belief there was a number of victims 'out there' who were in need but were not receiving any disaster assistance. Mental health counseling agencies functioned as an important adjunct of local programs that addressed unmet material needs.

Crisis counseling and emotional recovery

FEMA support for mental health services is an established part of federal assistance and emergency funding for crisis counseling was provided by FEMA within two weeks of the main shock. FEMA issued grants totaling $35 million to LA and Ventura County Departments of Mental Health (DMH). The Ventura County DMH received funding to support more than one year of crisis counseling and outreach activities for people with emotional needs related to the disaster and recovery. Project COPE (Counseling Ordinary People in Emergencies) was a systematic long-term crisis counseling program available across the county. (A similar program, Project REBOUND was offered in Los Angeles County.) Targeted populations included older adults, children, farmworkers, monolingual ethnic populations, homeless persons, and the unemployed.

To access at-risk groups, COPE subcontracted services to a number of local non-profit organizations that operated in the county on a regular basis. For children distressed by the earthquake, primary services were provided by INTERFACE, a Simi Valley CBO, and CAAN (Child Abuse and Neglect) in Fillmore and Piru. Work with children used schools as a center for outreach and counseling services. Children also provided connections to families having difficulties coping with the social and emotional disruptions of the earthquake. Home-based or clinic-based counseling would be used to follow up on school-based interventions if warranted by family problems. Outreach and crisis counseling for elder populations was conducted by Project COPE staff, and by OASIS. Counseling was provided at Senior Centers in Simi Valley and Fillmore, with additional services at mobile home parks and home-based visits for disabled and other homebound elders (Ventura County Department of Mental Health 1995).

Farmworkers, including many monolingual Mexican immigrants, were a harder population to access, given the complicating issues of residency and immigration status. To provide culturally sensitive Spanish-language services, COPE contracted with Catholic Charities and Clínicas del Camino Real. Both organizations had a long history of providing social and health services to poor Latino farmworkers in Ventura County and thus were trusted by victims otherwise uneasy with 'official' disaster assistance. As in LA County, Catholic Charities provided services to immigrants, irrespective of their legal status, including counseling for stress as well as referrals for material and financial assistance. Some of the Mexicans served by Clínicas and Catholic Charities reportedly had been victims of the 1985 Mexico City earthquake, and were easily distressed by yet another destructive earthquake. Farmworkers were familiar with Clínicas for health care services, making the provision of emotional counseling more easily accomplished. For legal immigrants, Catholic Charities took a strong advocacy role in securing their low income clients federal and state entitlements.

The cultural and demographic range of earthquake victims demanded a diverse set of locally knowledgeable organizations and a range of counseling and outreach strategies. ESCs in both LA and Ventura counties were staffed with REBOUND and COPE counselors to provide immediate assistance at the centers. For the 680 households in Ventura County that had to relocate, COPE staff provided emotional counseling services as part of local housing authority efforts to place families in new housing. For elders, focus group sessions at mobile home parks and Senior Centers provided the camaraderie and social support that relieved individual anxieties. In many cases, counselors provided both the emotional support and logistical services needed in obtaining relief services. As significant stress sometimes accompanied acquiring assistance and rebuilding homes, crisis counselors often provided their clients with transportation to the ESCs, housing information, and help in filling out application forms as part of their support services (Ventura County Department of Mental Health 1995).

Project COPE and REBOUND counselors referred to the changing emotional problems and needs clients experienced as they attempted to recover. Early crisis counseling focused on assisting those with fears associated with the trauma of the earthquake and coming to terms with what had occurred. Since their lives had been disrupted, people often needed to go through what was described as a 'grieving process for what they had lost'. Over time, as people attempted to acquire federal assistance or get adequate insurance settlements, some clients became disillusioned and frustrated.

COPE counselors described periods of anger victims would experience, directed both at an 'impersonal bureaucracy' and at themselves for the loss of control over their lives. Recovering emotionally from the earthquake and the recovery process took much more time than many disaster victims had expected. One year after the earthquake, a Santa Clarita crisis counselor described her own continuing fears:

> How have I recovered? Well, I put up three pictures. Okay, that's all I've done. I don't buy glass knick-knacks anymore. However I have the house is fine with me. I tell myself, 'You've got to do this. You've got to go out to the garage and straighten out the boxes and see what you need and don't need.' But, as soon as I open those boxes and take things out, I just snap. I put things back into the box, and I close it up . . . [whispering] Why? I don't know. I don't know why. That is the one thing I just can't deal with.

FEMA funding for the earthquake counseling services ended in May 1995. Among the agency representatives there seemed to be a general agreement that funding and length of time for emotional assistance was adequate for most victims' needs. Most disaster-related distress is transitory, but with earthquakes the persistence of strong aftershocks lasting a period of months can produce recurring fears and anxieties, particularly among children. In addition, for marginal households with continuing difficulties in acquiring assistance and rebuilding, situational distress can continue well beyond the initial shocks of the earthquake itself (e.g. Bolin 1994a). Project COPE provided counseling and referral services to nearly 45,000 in Ventura County. These included 8,500 elders, and more than 22,500 children, with approximately three times as many females as males using project services. While the project was directed at mental health issues, it also functioned to assist people with unmet needs by providing information, practical assistance, and referrals to NGOs for recovery assistance.

GENERAL PATTERNS OF VULNERABILITY

Drawing from the case studies and evidence from the City of Los Angeles, we return to issues of vulnerability in the context of First World societies (Figure 2.1). The diversity of experiences after Northridge is evidence of both a well-developed government-based system of social protection and the persistence of vulnerability among certain groups. Vulnerability in this case had little to do with systematic differences in exposure to the earthquake. While lower income groups frequently lived in older, less safe structures, earthquake losses were proportional across structures of all ages and values (Comerio 1996). As seen in Simi Valley, some middle class neighborhoods encountered significant damage problems because of localized soil conditions that continued to crack foundations for two years after the main shock. But vulnerability was revealed in differences in access to resources to manage losses, and those differences generally related to existing patterns of social inequality.

Persistent and situational vulnerability

FEMA and OES made extensive efforts to provide assistance to a regional population heavily diversified by age, ethnicity, national origin, language, and social class. The relief effort, whether motivated by political or humanitarian concerns, may be judged a significant improvement over previous efforts in which low income and minority groups were routinely overlooked or excluded from assistance (e.g. Peacock *et al*. 1998; Schulte 1991). Even in cases of US disasters when marginal groups are excluded from assistance, there is some level of public accountability for the state's actions, an accountability that is often lacking in Third World disasters (e.g. Oliver-Smith 1994). Nevertheless, statutory limitations on programmatic support, variations in settlement amounts, and cultural and political factors inhibiting access can produce or reinforce inequities in the relief process.

For vulnerable households, disaster-related needs concatenated with already existing needs, deepening the disadvantages people faced. Most federal disaster programs proceed from the assumption that people's problems begin with the disaster; therefore only those disaster-related problems are intended to be addressed by the assistance. By international standards, US federal programs are generous, but they nevertheless typically provide only partial compensation, and victims often must find other means to recover losses. From a vulnerability perspective, the issue is how existing disadvantages increase people's susceptibility to disaster and negatively impact their access to resources to cope with the aftermath. The evidence reviewed suggests that combinations of factors in specific social and geographical contexts limited people's response capabilities, and these limitations were not overcome by federal programs. The prevalence of unmet needs can be taken as an indicator of pre-existing social and economic deficits on the one hand, and limited access to resources, on the other.

At some risk of simplifying historical complexity, a distinction can be drawn between persistent (or chronic) vulnerability and situational vulnerability. Persistent vulnerability connects with structural factors, which often have a long history and which produce economically, ethnically, and culturally marginalized groups, usually with gendered variations within categories. Situational vulnerability occurs when fractions of a normally secure group, due to a combination of circumstances, become increasingly at risk in the face of calamity. The circumstances may be personal matters such as illness or accident, or they may relate to the recent restructuring of the political economy, which has produced rising levels of middle-class unemployment in the US. Situational vulnerability may result from bad luck or timing, and it cannot be as easily identified as vulnerability rooted in more persistent structural disadvantage. Of course, if situationally vulnerable households are not able to regain their former security due to continuing unemployment or underemployment, they may 'move' into the persistently vulnerable category as temporary disadvantage becomes chronic.

Interviews with FEMA, OES, and ARC personnel engaged in unmet needs activities suggest that several factors combined to produce situational vulnerability as seen in the suburban communities examined. These included declining market values of homes after a 1980s housing boom, job layoffs as a result of corporate restructuring and defense spending cuts, increasing indebtedness from underemployment, and high deductibles on earthquake insurance. For example in the last decade Los Angeles County has lost more than 100,000 well-paying aerospace/defense related jobs, with loss of possibly twice that many in support industries as a result of a multiplier effect (Soja and Scott 1996: 14). As an OES worker from Santa Clarita described:

> Some people around here are in a bad way because they can't afford their insurance deductible. One person applying for assistance here [the ESC] had $187,000 in damage and she got $2,700 from her insurance company! She was trying to get an SBA loan but that'll take awhile . . . The problem is, layoffs in aerospace have hit the town pretty hard and a number of homeowners were already in trouble on their mortgages. Without jobs people can't get SBA loans so they can only qualify for temporary housing assistance from FEMA, but they can't use that to fix their houses. And FEMA won't give mortgage assistance if they are behind on their [mortgage] payments.

Middle-class homeowners heavily indebted and without current employment were unable to find the resources to rebuild homes that, prior to the earthquake, might have had a market value of $300,000, and bank foreclosures followed. Even maximum IFG money ($22,200) was seldom enough to repair even moderately damaged homes. While unmet needs programs from the ARC, and NGOs such as Habitat, could provide some assistance, their programs were generally designed for lower income victims with modest housing losses. The continuing assistance program of the ARC, could offer grants of $10,000 only for those who had received the maximum IFG award, but in the first year it had processed fewer than 200 such cases. While situational vulnerability may affect only a small proportion of middle-class homeowners, the issue bears watching in future disasters as economic restructuring, deindustrialization, and corporate downsizing create changing financial and employment situations for increasing numbers of middle-class households (e.g. Soja 1996b).

The compounding of marginalization factors in the study communities – low income, Latino, monolingual, immigrant, unemployed, farmworker, substandard housing, elderly, social isolation – in various combinations marked persistently vulnerable households at the study sites.[12] In this sense, they were little different from marginal groups caught up in the Loma Prieta earthquake of 1989 (Phillips 1993), where lower income farmworkers, ethnic minorities, recent immigrants, elders, and homeless persons encountered housing and recovery problems. Losses resulting from Northridge and an inability to acquire adequate federal assistance meant that some victims faced long-term deterioration in their living conditions. The second tier of social protection offered by CBOs, NGOs, and social service agencies, did provide additional resources. However, undocumented immigrants, intentionally disadvantaged by federal policy, were necessarily limited in their ability even to use this second level, remaining largely invisible and uncounted.

At the smaller sites, recent Mexican immigrants were reluctant to apply for federal assistance to which they were legally entitled. Based on their life experiences in Mexican disasters, many immigrants did not expect the federal government to deliver relief assistance services, nor did they attempt to obtain any. Of a dozen interviews with Mexican immigrant workers in Fillmore, only one household applied for any federal assistance. As one victim explained:

> Well, maybe if [the earthquake] had been harder on me, I would have investigated it more. I heard that there were many people applying who really didn't need it, but I didn't go myself because I didn't need much. Thanks to God, the house didn't fall, and here we were. At least we had a roof over our heads. We had water, we always had canned food in the house, so we were fine. I just didn't go [for assistance] because I said to myself, 'There are many people who really need it, the ones who have lost their homes, lost their possessions.' I said, 'God will punish me if I am taking from those people because, thanks to God, I still have my house. It will be a sin if I take blankets from those who need them because I still have blankets in my home.' [translated from Spanish].

As described previously, the shifting federal policies to deny immigrants disaster assistance undermined the confidence of both illegal and legal immigrants. A Ventura social service agency representative reported:

> Even legal residents were afraid to go file for assistance or ask for services, scared that they would be deported. It didn't matter whether they were illegal or legal. They still had to be convinced that

this was their right. They had paid their taxes, and they were entitled to relief services.

In Fillmore and Piru, where a substantial undocumented migrant population resides, interviews conducted with immigrants[13] suggested that while many took advantage of the emergency relief available, including food, water, and clothing from churches and the local Red Cross, they avoided the DACs. When questioned about undocumented immigrants seeking assistance at the DACs, an agricultural worker from Fillmore told us:

> [It was] mainly the undocumented [workers] who don't have papers, they didn't even apply. They were the ones who were really affected. If you don't have papers and you go to the center, they ask you for identification. You don't have a driver's license, no identification, no papers. Those people are still living in houses that were really affected by the earthquake, but because they're scared of the authorities, they are not going to go and apply for benefits. I know maybe four or five who didn't go out of fear. Many people, even those who are working on their papers, applying for residency, they didn't go because someone told them that applying for assistance would hurt their residency application.

Agency respondents at all four sites recognized the covert nature of the immigrant population and the inherent problems in assisting undocumented foreign nationals under current political conditions.

In Simi Valley and Santa Clarita, elders living on fixed incomes were among the persistently vulnerable and prone to be disadvantaged in recovery. Those who had pre-existing contacts with social service agencies through 'senior centers' and related institutional supports typically received the assistance needed. The importance of established social support networks was significant in buffering the disaster's effects and in promoting the acquisition of recovery resources. Agency informants identified isolated elder households in the larger communities who had still not accessed any assistance even two years after the earthquake. In reference to her own neighborhood, one retired Simi Valley woman commented:

> The people I feel sorry for are the seniors and the handicapped. There are a lot of low income people who don't have the education who are still not getting help. They are still sitting there, living in unsafe houses. The city comes and tells them, 'You are going to have to repair this wall or take down this chimney.' Well, they don't have the money, so they don't do the repairs. What will happen to them the next time that we have a major earthquake?

Among older residents, those who are physically frail or disabled have fewer physical defenses against disaster than others, and it is recognized that they will be more dependent on service providers in the case of disaster. Through the activities of Area Agencies on Aging, elders did have access to additional protective measures, including health care, food delivery, and other support services. The particular concern of service providers was with elderly households outside this system of social protection and the defenses it offers against the physical and economic effects of disaster.

Vulnerability, political culture, and everyday life

These various elements of vulnerability derive not from the hazard agent or the built environment, but rather from how those exposure factors intersect with the conditions under which people live their day-to-day lives. To be sure, had the earthquake not occurred our respondents' lives might have been very different on 18 January 1994. Nevertheless, even without the earthquake, their vulnerabilities would have persisted, connected to their lack of economic or other resource entitlements, their political disenfranchisement, and related constraints on their capacities to manage their lives (e.g. Hewitt 1997). While the elaborate infrastructure of social protection put in place after Northridge buffered the effects of disaster across a widely diversified population, the underlying sources of vulnerability were largely unaffected. This is not to suggest that the state interventions were not valuable, but rather only that they make proportionately more resources available to more secure households, and they do not, by design, change underlying conditions.

As examples from the case studies suggest, those already marginalized and economically exploited often found themselves in the second tier of social protection, seeking assistance from the local programs and social services. For many marginal households at the study sites, these community-based programs worked to stabilize their housing situations and promote recovery over the two years they were in operation. As these organizations used local knowledge and grassroots participation to develop and deliver services, they were able to assist vulnerable households that had been denied adequate federal assistance. By relying on familiarity with local conditions, and drawing on local community figures, they were able to work in a culturally diverse setting. In the next chapter we examine how some programs went beyond restoration and introduced longer-term development programs.

Last, the effects of the political culture and social policy directed toward undocumented immigrants were significant, if somewhat difficult to demonstrate. The ideological campaign being waged in the US variously against welfare programs, and *ipso facto*, against poor people of color who may use them, received its most clear expression in efforts to deny immigrants without legal residency any social, health, or educational services. The coupling of anti-immigrant policies with federal assistance to disaster victims was without precedent in 1994.The California political scene, by most accounts, added to the general significance of immigrant status (and ethnicity) and worked to inhibit victims' access to resources. Thus while the federal government, through the efforts of FEMA, carried out its largest and most costly disaster response to date, congressional ideologues attached policies to intentionally disadvantage a possibly large group of mostly very low income persons. Since Northridge, the federal policies on immigrants, welfare programs, and disaster assistance, have changed further, and potential implications of those changes are discussed in the following chapters.

NOTES

1 The term 'individual assistance' refers to all federal and state relief programs that are accessible by qualified individuals and households. It is distinguished from 'public assistance', which involves programs that serve the public sector, primarily local governments and qualified non-profit agencies. SBA disaster loans are generally not included under the category of FEMA individual assistance grants although they are an essential part of federal support for recovery. Data on HUD Section 8 Housing Vouchers, because they are not technically a disaster relief program, are missing from OES tabulations of individual assistance data (e.g. OES 1995, 1997).

2 The Modified Mercalli Intensity Scale is based on perceived intensity of ground shaking combined with observed levels of material damage from the ground shaking. MMI contours can be drawn through any area that has experienced a damaging earthquake connecting areas of equal damage. The scale ranges from I (not felt except under favorable circumstances) to XII (total damage to built structures) (Smith 1996). For Northridge the intensities MMI VI to IX cover the areas of damage (descriptions are abridged):

VI: Felt by all, furniture moved, damage to chimneys and plaster.

VII: Everyone runs outdoors, damage slight to well constructed buildings, considerable damage in badly designed structures, chimneys may collapse.

VIII: Damage considerable in ordinary substantial buildings, great damage in poorly built structures, walls and columns collapse.

IX: Damage considerable in specially designed structures, damage great in substantial buildings with partial collapse, buildings shifted off foundations, underground pipes broken.

3 The waiting list for Section 8 certificates is so long in Los Angeles that it is often closed to new applications for years at a time. In 1989 when new applications were accepted after a lengthy closure, 180,000 people telephoned to have their names added to the waiting list in a four-day period (Wolch 1996). When HUD suddenly offered 20,000 new certificates to Northridge disaster victims, thousands of applicants who had already been waiting years for subsidized housing were bypassed, raising significant equity issues.

4 The deductible on earthquake insurance is based on the value of the insured property not on the cost of the damages. An owner of a $300,000 house with a

10 per cent deductible would need to pay the first $30,000 of any earthquake claim above that amount. In areas of Los Angeles and Ventura counties, high property valuations resulted in predictably large deductibles. OES (1997) estimates insurance deductible payments amounted to nearly $4 billion, an amount more than one-third the value of total residential insurance settlements. An unknown portion of the deductible payments were met through SBA loans (Comerio 1996) suggesting that long-term indebtedness to repair earthquake damage was not prevented by having earthquake insurance.

5　Landlords do not have to accept HUD Section 8 certificates, which are a rent subsidy agreement between the tenant, landlord, and local county Housing Authority. They are not cash grants that can be used at the renter's discretion. Fair housing laws in the US *technically* prohibit landlords from refusing to accept those with FEMA temporary housing grants, although informants in LA CBOs indicated that ethnic discrimination did occur. The FEMA housing grants offer more flexibility in housing choice, but for what is, for renters, a far shorter period of assistance than HUD certificates.

6　In the US, government criteria maintain that normally a household should spend no more than one-third of its monthly income on housing rent or mortgage. Thus 'affordable housing' must meet that criterion for low income households. The availability of affordable rentals in Los Angeles was sharply reduced during the rapid escalation of rents during the 1980s, when average rents increased 50 per cent over the decade, while median incomes remained stagnant or decreased for lower income households. By 1990, 75 per cent of those living in poverty in Los Angeles were paying more than 50 per cent of their income on housing (Wolch 1990: 404). According to Wolch's research more than half of all renter households in LA cannot afford a one-bedroom unit and 60 per cent cannot afford a two-bedroom unit. The net consequence is overcrowding as low income households began to share housing units or live in small substandard units (Soja 1989). According to federal housing standards, moderate overcrowding is 1.0 to 1.5 persons per room, and severe overcrowding is anything in excess of 1.5 per room.

7　The term 'permanent' is relative at best. As part of the attack on federal social programs in the US, conservatives in the congress have introduced bills to dismantle HUD and severely constrain its few remaining housing support programs (GAO 1997). The effects of this intentional disadvantaging of marginal populations will be taken up in the concluding chapter.

8　Condominium complexes are frequently converted apartment complexes that have moved 'up market' from rentals to owner occupied units. Condominiums are owned by the purchaser but are in buildings that are managed by a homeowners' association to which each condominium owner pays an annual fee for insurance, maintenance, and use of common areas (grounds, swimming pools, recreational facilities, parking garages). Townhouse complexes are similar to condominiums except that the owner of the townhouse also owns the land on which the townhouse is built. Townhouse complexes typically also have homeowners' associations to manage the common areas. Common in Southern California are expensive housing developments known as 'gated' communities. These are usually upper middle-class (or higher) single-family housing developments that are walled and have restricted access through a gate. The common areas in such developments are also managed by homeowners' associations to which each household pays an annual fee. Homeowners' associations can be very active politically in maintaining communities and opposing developments that threaten property values – especially low income housing (Davis 1992; Wolch 1996).

9 Both CBOs and NGOs are in the general category of non-profit service oriented organizations, either with volunteer or paid staff. For our purposes, a CBO is a non-profit organization that is based locally and not part of a national or international organization. NGOs may have local branches that deal only with a given community, but nevertheless have a national organization to draw resources from, if need be. After Northridge some new CBOs were formed while other CBOs extended their service domains into the realm of disaster assistance. While disaster NGOs typically have more expertise with national and international systems of disaster relief, CBOs have more local contacts and local knowledge about immediate needs and conditions.

10 These figures were provided by the Fillmore Building Department after its damage survey and are somewhat larger than those presented in Table 5.6 using OES data. The social construction of disasters through official statistics is not without its ambiguities and inconsistencies.

11 A Specific Plan is a generic term used for detailed planning documents that are employed by Redevelopment Agencies in California when they undertake economic development programs as part of an overall General Plan that each municipality is required to have under California code. The Specific Plan for Fillmore was completed the week of the earthquake and had to be amended to take into account the damage to the downtown area. The Specific Plan was produced by an architectural firm contracted for that purpose and lays out in detail land use planning, architectural design elements to be used, building uses allowed in different zones, as well as design of parks, streets, street lighting, and related infrastructure. The intent of the Fillmore Specific plan as adopted is to rebuild downtown Fillmore to capture its 'historical' character, to enhance urban amenities, and to expand commercial development and business activity. The old buildings in downtown Fillmore (and Piru) have been used as settings for various television and feature film productions and the Specific Plan seeks to reinvent the '1920s' look that has made it attractive to Hollywood film productions.

12 None of the social service or disaster agencies we worked with specifically referred to gender issues nor did any of our respondents. Only in the crisis counseling programs were women identified as needing specialized services. Our reading of the invisibility of gender to the service providers is that any differences were subsumed under general categories of poverty and age. Female-headed households among all ethnic groups have significantly higher poverty rates than other households. Similarly women, for both historic and demographic reasons, are over-represented among lower income elders. Yet social service agencies and CBOs assisting such households referred to poverty, ethnic discrimination, and/or age as key features, not gender. See Enarson and Morrow (1998) for a discussion of these issues after Hurricane Andrew.

13 To avoid creating undue political risks, none of the respondents was asked his or her residency status. Rather respondents were simply asked to tell the interviewer about the experiences of undocumented immigrants that they knew of.

Because the migrant population fluctuates and illegal immigrants attempt to remain invisible to the gaze of government agencies attempting to control and regulate, we do not offer any estimates of the numbers living in this area of Ventura County.

6

RESTRUCTURING AFTER NORTHRIDGE

Studies of recovery from disasters have documented the emergence of political and social conflict after disasters as people and groups seek opportunities for political empowerment and economic change (Olson and Drury 1997; Poniatowska 1995). Instances of these change dynamics emerged after Northridge, a process we refer to as restructuring. Because the institutionalized assistance system could not adequately deal with all circumstances, new social, organizational, and economic arrangements emerged. An important element in this process was the development of innovative programs for housing, preparedness, and social protection, both through local governmental agencies and the non-governmental organization sector. In the process of rebuilding following the earthquake, different groups, both inside and outside of existing power arrangements, mobilized to influence the agenda in post-disaster community reconstruction and development.

Unlike the Loma Prieta earthquake of 1989, where the federal government's position reinforced the peripheral status of already marginalized minority households and prompted political resistance mobilization (Bolin and Stanford 1991; Schulte 1991), the political situation after Northridge has been more open and fluid. Both FEMA and OES consciously addressed the Loma Prieta failures to assist ethnic minorities, and the federal government took an extended role in supporting recovery activities and promoting access to assistance. As a result, local governments, as well as NGOs and CBOs, were able to pursue projects that went beyond a mere return to the status quo ante. In various ways across the region, recovery, reconstruction, and development programs conflated, taking the earthquake as a point of departure, but in fact pursuing agenda to reshape communities, not merely repair the damage. Taken together, programs operating on different spatial scales constituted important elements in post-earthquake recovery and restructuring processes. While the major thrust of the chapter examines the ways the research communities engaged in redevelopment and economic restructuring, we begin with a review of housing and social protection issues in regional perspective, focusing on the City of Los Angeles and its efforts at maintaining neighborhoods and promoting affordable housing.

SOCIAL PROTECTION IN LOS ANGELES

At numerous points we have emphasized the high levels of social protection against disasters afforded by state, county, and city measures in California. While much of this is materialized in strict building codes and related engineered infrastructural protections, it is also instantiated in sophisticated emergency preparedness, well established linkages between state and federal emergency managers, political leverage to obtain federal resources, and technical expertise at managing disasters. These high levels of social protection cushion the effects of disaster on vulnerable households as reviewed in preceding chapters. The City of Los Angeles is at the leading edge of efforts to prepare for disaster and to manage recovery, and it is one of the only cities in the US to have a disaster Recovery and Reconstruction Plan (Tyler 1997b). Beyond the public sector, Los Angeles has a second level of social protection through a large network of NGOs and CBOs that offer services and resources to assist and protect the city's growing low income and immigrant populations (Chapter 5). In what follows, we discuss three areas of programmatic and organizational activity that attempt to address issues in social protection regarding both Northridge and anticipated future earthquakes. The first involves efforts by the city of Los Angeles to control residential loss and to maintain its stock of affordable housing. The second is the proactive networking of CBOs in the city and county to bolster response capabilities and offer enhanced social support services in LA's next disaster. The last concerns hazard insurance and the state's efforts at providing a joint public–private earthquake insurance to help fund reconstruction after the next destructive earthquake.

Residential abandonment

The earthquake was essentially a 'direct hit' on the vast suburb of the San Fernando Valley, which contains approximately half the city's population and has a high homeownership rate at 48 per cent (LAHD 1995). Three major concerns regarding the housing loss were the concentration of multi-family housing losses, the particular problems of low income populations, and funding the reconstruction of marginal housing stock. While these are easily categorized as recovery activities, because they are attempting to manage a longer-term structural condition in housing stock and its growing mismatch with people who need housing (e.g. Comerio et al. 1996), we discuss them as part of restructuring. In this sense, the disaster can be understood as a moment in a larger social process, one in which large-scale economic restructuring, changing federal housing policies, growing economic marginalization and polarization, and class-based politics, have conspired to produce current housing conditions in Los Angeles (Wolch 1996). These conditions include the largest homeless population in the US along with a surplus of 54,000 housing units at the time of the Northridge earthquake. In aggregate terms, the earthquake was a relatively minor event in contrast to the effects of two decades of economic restructuring and demographic change. Unlike the dynamic effects of restructuring that are continuously transforming LA, the earthquake's effects were compressed in time and they were reflexively managed by the public sector.

Los Angeles faced particular problems with areas of concentrated damage scattered across the city. Some 38 census tracts in the city (out of 756 total) had at least 100 vacated residential units, representing 12 per cent of the housing stock in them, and those 38 tracts contained two-thirds of the vacated housing in the city (Stallings 1996). The 15 most heavily damaged neighborhoods came to be called 'ghost towns', defined by the LAHD as census tracts containing 100 or more dwelling units with at least $5,000 damage per unit. In the ghost town neighborhoods an average of 60 per cent of the housing stock was severely damaged and vacated, and six San Fernando Valley neighborhoods lost between 80 per cent and 97 per cent of their housing stock (LAHD 1995). While in terms of total city and county housing stock, the losses were minor, its concentrated nature created many localized disasters.

For the city of Los Angeles, the overall loss of housing stock was less important than the localized social effects that neighborhood abandonment produced on remaining housing in the areas and on adjacent neighborhoods and businesses. As most of the abandoned structures were unsafe for entry, former occupants had to leave their belongings behind, and these became a target for thieves and looters. The concentrated areas of red tagged apartments and condominiums rapidly became blighted, as they attracted urban squatters, street gang activity, and crime. The seriousness of the decay escalated quickly, rippling outward toward intact neighborhoods on the peripheries of the ghost towns, a process which, if unchecked, could have led to abandonment of intact neighborhoods. The Los Angeles Police Department did not have the staff available or resources to put 24-hour patrols on the premises of so many abandoned areas. To enhance security, FEMA and OES committed $6.5 million in funding to provide private guards, debris clean-up, and fencing to control unauthorized access (FEMA 1995). Protection and security were only the first steps towards eventual rehabilitation and reoccupancy.

The ghost town neighborhoods were in census tracts that spanned a range of income and race/ethnic groups, from middle income Anglo areas to lower income and poverty class neighborhoods. Three ghost towns had a majority of Latino and African-American residents. While most of the lost housing was rental property, condominiums and single-family dwellings (rented and owned) were also included in the ghost town neighborhoods. Comerio (1996) characterizes the fifteen ghost town neighborhoods as relatively heterogeneous, with only four having predominantly low income (below $25,000) 'non-white' populations. Three ghost towns had median household incomes over $50,000, and eight had median incomes in the $30,000–$40,000 range. (Figures are based on 1990 census data for the tracts included in the ghost town areas.) Not surprisingly, the three ghost towns with higher income populations had a greater percentage of owners (mostly of condominiums) and a lower percentage of older rent-stabilized rental units. Conversely, the four lowest income ghost town areas had median household incomes ranging from $18,000 to $24,000, with more than 90 per cent living in rental units, and most of the older housing stock coming under LA's rent control ordinance. The one exception was the mostly African-American and Latino 'West Adams' area of South Central LA which, in spite a median income of only $21,000 – lower than Piru's – had a homeownership rate equal to the San Fernando Valley's at 48 per cent of all residential units. The four poorest ghost towns had resident populations that ranged from 31 to 94 per cent 'non-white' at the time of the earthquake (Comerio 1996). In December, 1994 the LAHD added two more neighborhoods to its list of ghost towns, with 628 and 43 vacant units respectively. This designation made special ghost town funding available to owners in the areas and brought the total number of ghost towns to seventeen.

The major concern of city officials was to get abandoned buildings refinanced, reconstructed and occupied. The LAHD, using its extensive building ownership database, proactively sought out owners of damaged apartments to ensure timely movement toward rebuilding or razing of damaged structures. In some instances, the LAHD assisted in the sale of damaged buildings to non-profit housing corporations to promote affordable housing availability. As more than 48,000 apartment units needed rehabilitation or reconstruction, at an average cost of $43,300 for a red tagged unit and $9,600 for a yellow tagged unit (OES 1995), the costs of repair and replacement were significant. Comerio (1996) estimates multi-family housing replacement costs at $1,575 million for the city of Los Angeles. While this figure is for all damaged multi-family housing in the city, two-thirds of the damaged rental stock was in the ghost towns and the immediately adjacent areas.

The marginal economic situation of many of the apartment owners created significant difficulties in refinancing, and overall surplus housing and the 'soft' rental market added to the questionable economic viability of many of the housing complexes. Apartment owners with negative cash flow could not easily qualify for commercial loans or low interest SBA loans. Some owners had no incentive to rebuild, since it would be potentially difficult to recoup reconstruction costs, particularly given the new more expensive seismic building code standards. For owners of apartments that were rent stabilized, recovering construction costs was even more tenuous. Specifically, apartments built prior to 1978 (including more than half of all apartments in the San Fernando Valley) are under the city's Rent Stabilization Ordinance. The median income of residents in rent-stabilized housing in the Valley is barely above the poverty line at $17,500 (Comerio 1996). While not all occupants of rent-stabilized housing are low income, this housing tends to be occupied by residents with incomes considerably below the Los Angeles median.

Mindful of the loss of low income housing stock after Loma Prieta due to inadequate funding, federal agencies and the city of Los Angeles engaged a variety of strategies to facilitate repair and reconstruction of apartments. Unlike Loma Prieta, low income renters were not disproportionately affected by the Northridge earthquake. Of all damaged units, 15 per cent had poverty level occupants, a rate directly proportional to the overall poverty rate in LA County (OES 1997). As noted in Chapter 2, the poor in such contexts are vulnerable not because of increased exposure to hazards but because they have limited access to resources to cope in the aftermath of disaster. In a city with high housing costs and a shortage of low income housing, housing choices are sharply constrained for those living in poverty. Even minor losses to the low income housing stock are far more significant than equivalent losses to housing stock for middle income residents, given their typically greater mobility and housing options. The situation is particularly critical in Los Angeles because of growing numbers of lower income persons and steady declines in low income housing stock due to gentrification or demolition (Wolch 1996). The 10,700 low income housing units damaged in the earthquake represents 2.3 per cent of the total low income housing stock (464,000 units) in Los Angeles County (Comerio 1996). Unlike rental housing for higher income residents, the housing market for low income persons is very tight, with vacancy rates under 2 per cent, essentially full occupancy given normal turnover rates (Wolch 1996).

In situations, such as that following Loma Prieta, owners of low income rental housing have no funding options beyond the SBA, which they cannot qualify for, and their properties remain unrepaired and ultimately foreclosed by mortgage holders. While the SBA is the 'normal' lender of choice to repair uninsured housing, owners of larger properties with marginal incomes are unlikely to qualify for its loans. Specifically, there were several heavily damaged large (600–1,000 unit) apartment complexes in ghost towns, and the SBA loan cap of $1.5 million for commercial disaster loans did not offer adequate capital to fund repairs (Comerio 1996). The LAHD and HUD anticipated such problems and developed programs to assist owners of rental housing. SBA also acted proactively to arrange funding for reconstruction by organizing a federal outreach effort to contact apartment and condominium owners and explain assistance packages available. As in Ventura County, the local housing assistance programs in Los Angeles were funded by special grants from HUD and were available to those property owners who had been refused SBA loans. Without these targeted funding programs an estimated 25 per cent of the damaged low income housing was in danger of not being repaired due to the precarious financial situation of the owners (LAHD 1995).

189

The HUD Housing Earthquake Loan Program (HELP) was a Northridge program that offered low interest loans to owners of damaged apartment complexes who were unable to secure other financing from commercial lenders (FEMA 1995). Funded at the $100 million level, HELP targeted loans to multi-family housing projects that already received other forms of HUD assistance (primarily mortgage insurance and project subsidies). Monies could be used for reconstruction of buildings as well as other types of assistance for tenants. It was used almost exclusively by owners of large apartment buildings in Los Angeles and the city of Santa Monica (Galster *et al*. 1995). To participate in the program, owners had to make one-third of all units in the reconstructed building affordable to those with HUD Section 8 certificates. Thus, the loan program was a tool to pursue agency objectives by incrementally increasing the availability of low income housing stock in the area, and it helped prevent the abandonment of low income apartment complexes (LAHD 1995).

The LAHD, using the city's considerable political leverage, lobbied for and obtained more than $321 million in HUD special appropriations targeting lower income housing rehabilitation and ghost town restoration (Table 6.1). The loans were financed through four different HUD assistance programs intended for affordable housing and community redevelopment. As part of the special Northridge appropriation (Chapter 3) more than $202 million (of $250 million total) in CDBG and HOME Investment Partnership funds from HUD were made immediately available to Los Angeles to fund a variety of programs related to housing needs. HUD waived numerous statutory requirements for the grants, giving the city considerable flexibility in usage of the funds. A special allocation of $22 million in HOME monies from the US president's discretionary fund was given specifically for rehabilitation of multi-family housing in designated ghost towns. Ghost town restoration was pursued with considerable vigor by the LAHD in the two years following the earthquake to promote neighborhood retention and stability (Stallings 1996).

Special loans were made available through LAHD's Earthquake Emergency Loan Programs to encourage rental unit owners to rebuild. Using $80 million of its HOME and CDBG allocations, the city developed the Acquisition, Reconstruction, and Rehabilitation Program and the Multi-Family Rental Property Program (LAHD 1995). The first targeted those who had, or were intending to purchase, damaged apartments to rehabilitate and rent out. In several instances this program allowed non-profit housing corporations to purchase apartments and develop new affordable housing units for low income renters. The second program was intended for original owners of apartments who could not obtain other funding for reconstruction. Loans were available for $35,000 per unit, up to $525,000 per apartment complex, and 15 per cent of the loan was offered as a grant to offset high labor costs. Both programs required that at least 20 per cent of the rehabilitated units in

Table 6.1 HUD funds allocated to the City of Los Angeles

HUD program	Total allocation	Available for loans	Rebuilding targets
HOME IP Supplemental	$58,995,000	$53,353,600	Single and multi-family housing
HOME IP Presidential	$22,467,000	$22,467,000	Ghost Town Apartments Only
CDBG Supplemental-A	$40,209,400	$40,209,400	Single and multi-family housing
CDBG Supplemental-B	$200,000,000	$195,000,000	Apartment Complexes Citywide
Total	$321,671,400	$311,029,800	

Source: LAHD (1995)
Notes
Reduced amounts available from loans due to administrative costs.
HOME IP – HOME Investment Partnership Funds.
CDBG – Community Development Block Grants.

apartments with five or more units meet low income affordability criteria. Together these programs issued more than 4,000 interest-free loans, with all payments deferred for six years, and made funding available to rebuild an estimated 15,000 damaged apartment units (LAHD 1995). A similar loan program was available from the LAHD to help finance the reconstruction of uninsured condominium complexes in the city, although no affordability criteria had to be met and interest rates could float to 7 per cent. Finally, in addition to the LAHD effort, SBA also issued more than 1,500 loans to LA apartment owners totaling more than $78 million.

In general, the refunding and reconstruction of multi-family buildings (apartments and condominiums) takes considerably longer than that of single-family dwellings, particularly if the latter are adequately insured. Recent evidence from California earthquakes and other US disasters suggest that the majority of single-family units are repaired and occupied within two years of impact (Comerio *et al.* 1996; Peacock *et al.* 1998). However, as a result of the concerted efforts by the LAHD, working with HUD funding, within one year funding had been arranged for 85 per cent of the abandoned apartment and condominium buildings in the ghost towns and reconstruction had begun on approximately one-third of them, and by end of the second year more than two-thirds of the units in the 'ghost towns' had been rebuilt and reoccupied (Comerio 1997). By the end of 1996, LAHD estimated that fewer than 5,000 red and yellow tagged units had yet to be repaired (10 per cent of all damaged units) and only 500 had been razed. However, as Mittler and Stallings (1997) note, while the ghost towns have been largely rebuilt, the neighborhoods themselves have been transformed through significant shifts in resident populations. This includes the creation of new ethnic enclaves in some areas, as former residents left the area to be replaced by internal migrants from elsewhere in the city. Thus a significant restructuring in home-ownership, business ownership, and the ethnic make-up of neighborhoods has characterized the recovery process. Given the dynamic shifts in the ethnic composition of neighborhoods across Los Angeles, such transformations are expected, and in this case were only accelerated due to earthquake-related population shifts from displaced victims.

Some have questioned the disproportionate attention lower income ghost towns received by the LAHD and the more than $300 million spent on the comparatively small number of victims in these areas (Mittler and Stallings 1997). Yet the prevention of neighborhood blight and restoration of rental housing are important urban housing goals, particularly for vulnerable lower income households. In that sense, the attention given by LAHD to the restoration of rentals must be seen within the frame of a larger housing crisis in Los Angeles over the availability of affordable housing. While Los Angeles becomes increasingly diversified by ethnicity, it also has a growing lower income population with larger than average households. As discussed in Chapter 5, LA's rental housing stock is increasingly poorly matched to the household characteristics of those who use it, both in terms of size and cost (Comerio 1996). From this perspective, any effort to maintain or increase affordable housing stock is valuable, especially given the increasing problems of overcrowding and homelessness (Wolch 1996). Lower income renters in LA are paying an increasing share of their incomes for housing and are slowly getting priced out of the market. As a result, tenants increasingly share the rent burden with others leading to severely overcrowded apartments (Soja 1989).

The ghost town/apartment issue also demonstrates that market mechanisms do not work to provide or rehabilitate affordable housing for an urban low income population. In this case, government interventions were necessary to manage a housing problem made worse by the earthquake. Ironically, the shortage of affordable housing in Los Angeles can be attributed, in part, to the cessation of HUD funded federal housing construction in the 1980s along with the city's encouragement of gentrification (Wolch 1996). The freeze in the construction of new public housing, combined with increases in the numbers of low income households needing housing assistance, suggests the housing shortage will only worsen. While Section 8 certificates were a partial solution after Northridge, they do not provide a consistent demand and thus do not lead to the creation of new housing. Major cutbacks in other federal and state welfare programs in the 1990s have worsened the affordability crisis in housing in Los Angeles (Wolch 1995).

The 'solutions' to multi-family housing reconstruction after Northridge thus fell to a rapidly assembled patchwork of loan programs, flexible grants, and outreach efforts primarily by federal agencies and local municipalities, most notably the City of Los Angeles. SBA was an important lender for smaller (2–9 unit) apartments damaged in the earthquake, as it was for single-family dwellings (Comerio 1996), though it is poorly adapted to the needs of owners of larger economically marginal complexes. HUD, through its own loan program, and the LAHD became key players in getting apartment complexes, particularly those in blighted areas, funded and repaired. In so doing, the LAHD was able to pursue its own agency objectives by both restoring lower income housing, and adding units of affordable housing to the existing stock. The problems confronted and efforts mounted in the city of Los Angeles to solve them represent a unique, if *ad hoc*, combination of funding opportunities and city expertise in promoting specific recovery and housing goals.

While the mosaic of programs and agencies has helped to reconstruct many of the damaged apartments (90 per cent of all vacated units were apartments), they can do little to address the underlying causes of the area's housing problems. The programs are necessarily limited in scope, funding, and duration. There are numerous unknowns as well. While buildings are rehabilitated, non-structural ('cosmetic') damage, and damage to parking facilities and appurtenant structures may remain unrepaired. Similarly green tagged apartments that remained occupied may forgo expensive repairs to non-structural damage. Thus there may be a deterioration in the quality of housing stock in high-damage areas, with its attendant effects on property values and neighborhood aesthetics (Comerio 1996). In addition, the long-term economic viability of affordable apartment complexes in light of loan pay-back requirements is unknown, particularly for those that had marginal profitability even before Northridge. As most LAHD loans had six-year payment moratoria, owners have yet to begin their payments. The reconstruction and building rehabilitation does not mean a return to a level of housing quality in existence prior to the disaster, however effective programs may be in preventing the loss of existing housing stock.

NGO Networks and vulnerable populations

As described in Chapter 5, non-profit social service organizations were important additions to social protection measures after Northridge. We characterized them as a 'second tier' of protection as they filled gaps in federal programs and dealt with excluded and marginalized populations. The primary vehicle to assist victims was the creation of several local and county-wide organizations to coordinate unmet needs activities in Los Angeles and Ventura Counties. These worked to systematically develop relief alternatives for those who failed to get resources commensurate with needs. Northridge had two important restructuring effects on the voluntary organizations arena. First, LA County NGOs and CBOs renewed their efforts to develop a framework of coordination for the next large urban disaster. Second, the increasingly exclusionary criteria of federal disaster programs generated significant new concerns about how the non-profit sector can operate in future disasters. These concerns may be compounded by significant reductions in federal funding, which could increasingly shift burdens of assisting victims onto the poorly funded NGO sector.

Los Angeles County already had its own county-wide Voluntary Organizations Active in Disaster (which complemented the Southern California VOAD). Yet the 1994 disaster had considerably broadened the field of involvement by CBOs in disaster beyond those in the LAVOAD. To develop an informational and inter-organizational network for all interested CBOs, in the months following the earthquake a new umbrella coordinating organization – Emergency Network Los Angeles (ENLA) – was developed at the behest of the mayor, and in 1995 it was formally established as a non-profit corporation. While ENLA, and its 300 member CBOs, was involved in aspects of Northridge recovery, much of its current activity is oriented toward networking for future disasters. It organized a Housing Task Force, which coordinated activities of member organizations and the Unmet Needs Committee, and focused much of its effort on assisting victims when housing benefits and Section 8 Certificates expired. It also assisted condominium owners and other homeowners' associations with securing funding to repair common property facilities.

The broad framework of ENLA, which now incorporates LAVOAD, is directed by an executive committee and divided into six operational regions across LA County. It has four major committees to address issues regarding housing, technology, information, and funding. Member CBOs are computer linked for the exchange of information, and periodic meetings (at least monthly) are held for participating organizations. While only registered non-profit organizations can be voting members, ENLA also includes interested government agencies, businesses, and elected officials as part of its network. Funding for ENLA comes primarily from private sources, and each member organization is responsible for its own funding. In future disasters, ENLA will coordinate activities of its members with the ARC and will maintain a liaison at the Emergency Operations Center. Most important will be the coordination of donated goods and services, resource sharing amongst CBOs, coordination with responding government agencies, and the early development of an Unmet Needs Committee to serve vulnerable populations. As such, ENLA should be seen as an important attempt to expand social protection during emergencies, and to assist those whose needs are not addressed through the formal assistance system.

There are mounting concerns in the non-profit sector regarding federal efforts to exclude undocumented immigrants from all disaster assistance and over the severe cutbacks in federal social welfare programs and their effects on vulnerable persons (Wallrich 1996). New federal regulations (as of July 1997) require agencies at all levels of government to report any undocumented immigrants to the Immigration and Naturalization Service. Under these regulations, undocumented immigrants are ineligible for any type of government assistance and are subject to deportation. (At the time of Northridge, the only restriction on assistance to undocumented immigrants was that they were ineligible for assistance beyond a ninety-day period.)

A more significant intentional disadvantaging of immigrants and other marginal groups may be read from the US 1996 Welfare Reform Act, which placed severe restrictions on the amount of federal monies available for health, nutrition, disability, housing, and education assistance. While this does not directly pertain to disaster relief, it has cut welfare resources to vulnerable populations including the homeless, those living in poverty, female-headed households, immigrants, elders, children, and the disabled (Ehrenreich 1997). Under the new federal approach, states receive block grants with which they administer the remains of welfare programs. Because the block grants are significantly smaller than existing social expenditures, the net consequence will be fewer resources and more time limits on benefits. In urban areas with higher rates of poverty, and higher proportions of the population on welfare assistance, recovery after disaster could be severely impaired (Collins 1997).

Non-profit organizations may be hard pressed to provide for what are likely to be large-scale unmet needs resulting from federal cutbacks. Thus the vulnerability of specific groups, formerly protected by welfare programs, will be increased. The prospects are even worse for immigrants (legal and otherwise), as some welfare programs intended for elders will not be available to legal immigrants if they were not of a qualifying age in 1996. All legal immigrants must now wait five years before applying for any welfare assistance. Exactly how these restrictions will affect future US disasters is unknown, but it is likely that vulnerability will be increased from the loss of entitlements, and recovery rendered more problematic. These changing federal rules have created significant uncertainties among CBOs regarding how they will assist people in future disasters, and what kind of public resources may or may not be available to draw from (Wallrich 1996). It seems apparent that once ARC shelters are shut down after a disaster, there may be a large number of households that are ineligible for any food or housing assistance from the major federal disaster responders (Collins 1997). Whether networks of CBOs such as ENLA, or area-wide Unmet Needs Committees, will be able to provide service at usual levels to those increasingly vulnerable to disaster remains to be tested. This neoliberal effort to 'discipline and punish' those segments of the population with precarious lives serves as a marker of the subversion of humanitarian concerns to the larger project of market discipline currently fashionable.

Insurance issues

California's experience with a decade of disaster raises significant questions about how mitigation and recovery can be funded. The convergence of federal assistance and large insurance payouts after Northridge provided a pool of resources to fund many reconstruction costs. The assemblage of assistance dollars and programs was a direct result of two factors: the particularities of political claims and interests mobilized in the days following the main shock, and lessons learned from Loma Prieta regarding inadequacies in assistance for poor and ethnic populations and for reconstruction of low income apartment complexes.

In order to provide a financial pool that would allow the state to be a 'lender of last resort', California attempted to fund its own program to address the special needs of low income home and apartment owners. Funded out of state surplus monies and run through the state's Housing and Community Development department, the California Disaster Assistance Program (CALDAP) was launched in 1990 (Comerio *et al.* 1996). It was to provide a ready pool of money to lend to victims of future disasters who could not qualify for FEMA or SBA assistance. In addition, the state developed the California Residential Earthquake Recovery Fund (CRER). CRER, which was funded by a small surcharge on insurance policies, provided up to $15,000 to help insured homeowners pay the typically large insurance deductibles. While the coverage was obtained through private insurers, the state guaranteed the funds.

CALDAP survived long enough to issue approximately $700,000 in loans after the 1992 Cape Mendocino and Landers/Big Bear earthquakes. CRER likewise issued 6,500 payments against claims in the two disasters, averaging approximately $7,000 each (Comerio *et al.* 1996). Since CALDAP was to be funded by state surpluses, the deepening recession produced a fiscal crisis in the state and the program quickly became insolvent. And, although 90 per cent of earthquake insured homeowners had paid into CRER, the program was repealed in 1993, after being in effect one year (Palm 1995). After Northridge, in order to generate revenues to place in the CALDAP fund, two strategies were tried, as discussed in Chapter 3. One was a small increase in sales tax (a similar strategy was used after Loma Prieta), the second, proposed by the governor, was the sale of bonds. In a move that is emblematic of the politics of taxes and public programs in California, neither was approved.

In the absence of a public willingness to share the costs of recovery through state funds, the federal government, particularly HUD, was left to socialize the costs of the reconstruction of economically marginal housing. California's failures to provide a fiscally sound resource pool to fund private reconstruction costs raises concerns about what most agree are inevitable major urban disasters in the future and how these will be paid for. Shrinking federal budgets portend a shift of spending burdens downward to the state level, as has occurred as a result of other program cutbacks. Thus far, there is a continuing strong federal interest in hazard mitigation, but there are moves to reduce federal resources available in disasters as part of the politics of the 'balanced budget.'

Although it failed to provide a sustainable resource base to function as a 'lender of last resort', California has recently moved to address problems associated with homeowner earthquake insurance. Single-family dwellings constituted a major portion of all rehabilitation costs, but only about 10 per cent of the most seriously damaged housing stock after Northridge (OES 1997). (Among other things, this suggests that loss estimate procedures may seriously undervalue the cost of repairing 'green tagged' structures in areas with high cost housing.) And while only about 35 per cent of all single-family dwellings had earthquake insurance, claims rose steadily for two years, with the majority of the more than $10 billion in insurance going to homeowners, much of it for repair of non-structural damage to houses (Comerio et al. 1996). The ostensible fiscal crisis that the large payouts produced for some insurers initiated moves to raise the rates on existing earthquake insurance policies and suspend the sale of new policies on the grounds that risks could not be adequately spread over all policy holders.

At the time of Northridge, California law required that insurance companies offering homeowner policies also must offer earthquake coverage to all policy holders. The state insurance commissioner estimated that 70 per cent of homeowner insurance companies had left California for this reason. Within a year following the earthquake, the insurance commissioner initiated an effort to remove the earthquake insurance requirement for insurers, and in April 1995, the state Senate Insurance Committee passed legislation to repeal the law. Insurance companies contended that if the law were not repealed, their alternative would be to leave California and not provide any homeowners' insurance within the state. Insurance industry specialists advocated some redistribution of insurance coverage costs from private companies towards a state-guaranteed risk pool, which the state has moved to provide.

In late 1996, the California Earthquake Authority (CEA) was created. This state program, in conjunction with private insurers, currently offers a state-guaranteed earthquake insurance policy to homeowners with existing hazards insurance. The CEA coverage is furnished through those insurance companies still operating in the state (more than half the insurance market is controlled by three companies, State Farm, Allstate, and Farmers), and funds are guaranteed by the state through a debt authority (Comerio et al. 1996). As the program is only in its first few months of operation at the time of this writing (September 1997), no judgement can be made as to its fiscal soundness. The state estimates that more than $10 billion should be available in its first twelve months of operation (slightly less than the total Northridge insurance payout).

This new public–private structure offers a 'no frills' earthquake policy to homeowners through their regular insurers. In the case of an earthquake, all payments to policy holders would be made directly from the CEA. The coverage is marginal at best: the deductible is set at 15 per cent of the value of the home (private policies seldom exceed 10 per cent) and only the house itself is covered (not appurtenant structures such as surrounding walls, garages, swimming pools, landscaping). Premiums will be set at approximately $3.30 per $1,000 coverage, resulting in annual premiums in the $600–$1,000 range for average-priced homes in Ventura or Los Angeles Counties (CDI 1997). The CEA fails at the outset to address a key issue for homeowners: the cost of the deductible. For example, owners of a $300,000 house with $50,000 earthquake damage would have to pay $45,000 in order to receive $5,000 from the CEA. As Comerio *et al*.'s (1996) research shows, SBA loans were often used by homeowners to pay for the insurance deductibles so they could utilize their insurance to repair their homes and the CEA is unlikely to change that pattern.

As a quasi-public agency the CEA is under a governing board of three top state officials and a ten-member advisory panel. The advisory panel is heavily weighted towards the private sector and comprises four insurance companies executives, two insurance agents, one building expert, a seismologist, and two members of the public. In response to this composition, consumer advocates have claimed that the new agency will be unduly influenced by the insurance industry on issues such as rates. There are no data yet available on public participation in this new program, but since it is more expensive and offers less coverage than regular pre-Northridge policies, it remains to be seen if coverage rates will increase notably. In addition, while this attempt at restructuring earthquake insurance toward a quasi-public guarantee is a small step toward generating a resource pool for private recovery, it does not address insurance or reconstruction funding for multi-family housing, which continues to be the largest single source of housing in LA County and the most poorly served by federal reconstruction financing. The CEA is a measure of social protection, but its terms will be unappealing to homeowners already heavily in debt, many of whom are likely to continue to 'take their chances' without earthquake insurance in a risky environment.

RECOVERY AND RESTRUCTURING IN LOCAL PERSPECTIVE

The nature of damage in an earthquake is such that recovery necessarily involves more than the repair and reconstruction of the built environment. The dynamic nature of political and economic processes at all spatial scales necessarily shape reconstruction and recovery within a changing framework of existing and emerging power relations. In the US context, given the varying degrees of autonomy at different political levels and variously influential political and economic interests, the particular directions of reconstruction and development are malleable and by no means overdetermined. History does not begin with the earthquake, and the disaster process, including reconstruction, is influenced by the existing uses of space and political economy of an area. Additionally, given the prominent role of federal resources in recovery and development, the statutory limitations placed on expenditures from different programs constrain how such resources can be used. Such limitations were pronounced in the case of FEMA's public assistance programs; local governments had to distinguish carefully between rebuilding public facilities and other projects that were outside of FEMA's funding mandate.[1] Nevertheless, employing their own political strategies, city authorities used reconstruction as a means to move communities in what they viewed as desirable directions, allocating different federal resources to different development ends. The line between recovery and broader restructuring or redevelopment goals is seldom clear, though local authorities operationalized this conceptual separation in their actions in order to maximize their use of federal resources.

In the four communities there were development projects that pre-dated the earthquake and remained largely unaltered afterward, save for changes in implementation plans and new building codes. There were also instances where the earthquake created new problems (or exacerbated existing ones) that were handled in ways that reflected existing political priorities. The LA Housing Department's efforts to restore neighborhoods and expand the stock of low income housing is a key example. But the disaster also created instabilities in existing social relations that occasionally moved communities in new directions. It made public issues out of existing but invisible or un-addressed problems, providing both the political impetus and the resources to begin to redress them. Among these were the shortage of affordable housing and farmworker housing in Ventura County, and revitalization of the older, low income Newhall area of Santa Clarita. Rebuilding after the earthquake constituted a process (still ongoing) in which a complex of existing political interests and new political actors competed to shape the directions in which communities developed. Efforts at change were constrained by resource limits, prevailing land use patterns, and existing power structures.

As part of the recovery process, local political authorities initiated discussions over what problems the earthquake revealed and what strategies should be used to address them. Civic groups and non-profit organizations entered into the public discourse as well. On occasion, interest groups re-introduced existing issues, taking advantage of the earthquake to renegotiate previously denied political access. Programs that emerged varied across the region, ranging from those closely tied to earthquake reconstruction to projects that required social and political change and redistribution of community resources. Whether projects were intended to address long-standing grievances or enhance economic opportunities for specific classes, most depended on the resource availability and the temporary opening of political agenda that followed the Northridge earthquake.

Federal resources and local actions

From the beginning, political leaders and emergency managers in the peripheral communities constructed the disaster as one in which they were competing with Los Angeles, with its well-organized, politically astute staff and leadership cadre, for a share of federal disaster relief funds. The rhetoric of Ventura County being 'overlooked' frequently appeared in interviews with county agencies (see Note 3). Nevertheless FEMA Public Assistance monies are based on documented disaster losses to public facilities and emergency management expenses, thus, technically, different municipalities did not 'compete' for the $4.5 billion in FEMA assistance. On the other hand, receipt of public assistance is an exceedingly complex, cumbersome, and lengthy administrative process that involves both claims-making and negotiation with FEMA as to what will or will not be reimbursed (e.g. OES 1997). Indeed by mid-1997, a number of public reconstruction projects in the study communities had yet to be reimbursed by FEMA's public assistance program. By most accounts, FEMA was particularly responsive to requests made by Los Angeles County municipalities, particularly the city of Los Angeles for assistance (Nigg 1997). If there was competition for federal resources, it involved the acquisition of HUD and Economic Development Administration (EDA) resources to be used in development projects and affordable housing. In such instances, knowledge of federal guidelines and programs was a distinct advantage, and in these matters, the city of Los Angeles has few peers. Nevertheless the Ventura County Board of Supervisors did have familiarity in dealing with FEMA during recent disaster declarations and likewise had HUD CDBGs in place at the time of the earthquake. While lacking the political influence and large staff of Los Angeles, the county did have successes in pursuing federal resources for both repair of public facilities and implementation of recovery and development projects.

The Ventura County government received a total of $28.5 million in federal and state grants, with most of the money coming from FEMA public assistance, CDBG funds, EDA grants, and State Historic Preservation Office grants.[2] For redevelopment projects and related earthquake recovery needs, the $6.8 million in HUD and EDA grants offered the greatest flexibility in spending. HUD block grants allowed the county to issue grants and loans to municipalities and NGOs/CBOs for assisting households with unmet needs. In March 1994, the county applied for a $2.5 million grant from the EDA to fund business recovery, economic development planning, and the creation of an economic development corporation to manage implementation. It initially received $716,000 to establish the planning and implementation infrastructure, and the remainder was granted for actual business development projects, most targeted for Fillmore. The county also successfully obtained early release of existing block grants to be used in various earthquake related programs. Shortly after the disaster, it received $2.8 million in block grants, and these were later supplemented by an additional $1.5 million disbursement. A supplemental $400,000 grant through HUD's HOME Investment Partnership for affordable housing program was also given to the county to be used in the construction of low income housing. Most CDBG monies went to Fillmore and Piru, channeled through Fillmore's and the county's Redevelopment Agencies (see p. 203 ff).

While CDBG funds normally are not intended for disaster reconstruction, HUD waived statutory requirements allowing county and city (including Los Angeles) programs to directly fund low income housing construction. Ventura County officials targeted the smaller communities with grants to be used in commercial rehabilitation, housing reconstruction, and infrastructural repairs. Fillmore received more than $1.5 million of the county's block grant funds to use in home rehabilitation projects and business redevelopment. Of these monies, $800,000 went to rehabilitation of damaged businesses in the central business district and another $250,000 went to pay for temporary business facilities while reconstruction was in progress. County officials also directed $575,000 in CDBG monies jointly to the Fillmore Redevelopment Agency and Mano a Mano, for low income home repairs and reconstruction. In December 1994, the State Historic Preservation Office awarded the county $380,000 to be used for reconstruction of historic buildings in Piru's small business district. As with virtually all the special grant programs, to qualify for county assistance, homeowners had first to be denied SBA loans.

In contrast to Fillmore and Piru, Simi Valley had its own existing CDBG program and consequently did not need to work through county authorities. This larger suburban city directly received $5.5 million in HUD block grants to fund earthquake programs for rebuilding homes and businesses of those denied SBA loans. By May 1994, the Simi Valley city council approved funding for a series of programs for lower income households including a mobile home repair program, a commercial rehabilitation loan program, a home rehabilitation loan program, and an affordable housing project. The city council also approved use of $50,000 in HUD funds to cover the costs of a further geological study of soil settling and liquefaction on the southeastern side of Simi Valley. In July 1994, the city council developed guidelines for a loan program targeting residents who had not received sufficient assistance through FEMA or SBA. Simi Valley residents with incomes of less than $39,000 could apply for grants up to $5,000 or for loans of up to $20,000 for uncompensated earthquake losses. The city council allocated $50,000 in federal funds to hire an earthquake recovery coordinator who would assist businesses in acquiring resources necessary for their recovery, either through federal or private sources. Unlike the smaller and older communities, Simi Valley – a relatively new city – engaged in no major redevelopment programs, choosing rather to address issues that pertained directly to earthquake damage and the difficulties faced by lower income households.

Of the four communities studied, Santa Clarita displayed the most political acumen in its search for outside funding. Its location in Los Angeles County worked to Santa Clarita's advantage, as did the city's familiarity with FEMA procedures gained during an earlier federal disaster declaration (Nigg 1997). Initially, Santa Clarita received $4.6 million in CDBG funds for community outreach and programs for those with unmet needs. City staff actively pursued other funding, and within six months of the earthquake they had obtained $1.5 million for low income housing rehabilitation through the California Department of Housing and Community Development. Using these funds, the city developed a loan program for low income homeowners who had exhausted other assistance sources and still had unmet needs. Santa Clarita received a total of the $14.4 million in HUD and EDA grants and – through its redevelopment agency – began directing those funds to recovery programs and revitalization of downtown Newhall.

In addition, Santa Clarita emergency managers actively pursued reimbursement claims under FEMA's public assistance program, saying that the delays in their reimbursement indicated that the city was a 'low priority' for FEMA.[3] The city reported spending $6.8 million in debris removal, $4.6 million rebuilding City Hall, $12 million on roads, and $5.3 million on the emergency operation center. Within the year, Santa Clarita had received a total of $16.5 million in FEMA assistance to repair public buildings and infrastructure, and totals have since risen to more than $26 million.

In summary, the four communities displayed different abilities in acquiring federal assistance, although all received adequate funding to initiate rehabilitation and recovery programs. We emphasize that both FEMA and HUD were considerably more liberal and flexible in their disbursements after Northridge than after previous disasters. Fillmore and Piru relied extensively on Ventura County staff both to acquire funds and to establish distribution programs. Simi Valley operated on its own, relying on the city council and staff to acquire FEMA public assistance and to administer its HUD funded programs. Santa Clarita, like the City of Los Angeles, showed the capacity to identify, pursue, and acquire additional federal funding sources. Santa Clarita was also particularly ambitious in pursuing timely reimbursements from FEMA. The use of federal funds to address development goals was facilitated through the legal authority of redevelopment agencies to manage the development process.

Redevelopment agencies

Under California legal statutes, county and municipal governments have the authority to designate specific areas, usually neighborhoods or business districts, as redevelopment zones. To plan, finance, and implement redevelopment projects, redevelopment agencies are established by local governments. The agencies are typically authorized to use a share of existing and future property taxes to support projects. As redevelopment projects are assumed to enhance property values, tax support for the projects should increase as projects are completed. Redevelopment agencies, as components of local governments are also eligible for HUD and EDA grants to promote low income housing and business enhancement projects. Santa Clarita and Fillmore had redevelopment agencies in place at the time of the earthquake. Fillmore's had been created to implement the renewal of that city's central business district as part of its new 'Downtown Specific Plan' adopted just days before the earthquake (City of Fillmore 1994). Santa Clarita's Redevelopment Agency had been in existence since 1989, though it had been largely moribund prior to the earthquake. Ventura County officials created a county-wide redevelopment agency as an institutional mechanism to assist Piru in recovering and in enhancing its business and residential facilities. In each case the agencies allowed local governments to pursue specific agenda for future growth.

Fillmore's downtown rehabilitation plan was intended as a comprehensive re-working of the central business district to improve access, increase business activity, and provide a uniform architectural style to maintain, or more properly, invent the 'traditional character' of the downtown as a *simulacrum* of a 1920s railroad town (cf. Baudrillard 1988). Historic preservation had been a concern for Fillmore and Piru, as both towns are used as sites for Hollywood productions and television series seeking an early twentieth-century ambience. The revenues gained from such productions led officials to propose having the motion picture industry help redesign Fillmore and Piru as movie sets, a postmodern twist to earthquake recovery. The earthquake complicated implementation of Fillmore's downtown redevelopment because more than a dozen historic URM buildings were severely damaged and red tagged. The Fillmore Hotel – a residential hotel for migrant workers and the only large multiple occupancy structure in the city – partially collapsed in the temblor and was razed within two weeks. To proceed with downtown reconstruction and redevelopment, other damaged buildings, including a Masonic Temple and the Victorian-era Towne Theater, were slated for demolition. However, the State Historic Preservation Office (SHPO) intervened to delay demolition of the theater and temple, pending inspection by a preservation manager to ascertain the possibility of rehabilitation (Tyler 1997a).

Redevelopment Agency planners as well as owners of businesses housed in the historic structures objected to the external interference by SHPO that produced delays in demolition and reconstruction. The demolition of historic structures is an ongoing issue in California earthquakes and resulted in protracted legal battles both after Loma Prieta and Whittier Narrows (e.g. Bolin 1994a). Owners of historic buildings often lack the resources to engage in the expensive restoration necessary to bring them up to standards required by current seismic codes, yet they are legally enjoined from razing buildings and selling the lots. In the case of the Fillmore Masonic Temple, the owners could not afford repairs and historic preservation groups were unable to organize financing for its rehabilitation; thus it was demolished three weeks after the main shock despite a legal challenge. However, the Redevelopment Agency did receive a $485,000 grant from SHPO to assist in the repair and seismic retrofitting of the Towne Theater. To prevent demolition, the City of Fillmore had to purchase the theater and organize the remainder of funding for its restoration through use of CDBG funds.

Apart from delays associated with historic preservation issues, the implementation of the plan for revitalization of the downtown was coeval with earthquake reconstruction. However, disputes emerged as a result of local organizations contesting elements of the redevelopment process. Fillmore passed a city ordinance requiring new or rebuilt downtown businesses to have retail components on the ground floor of all buildings. Clínicas del Camino Real, a non-profit farmworker medical clinic in downtown Fillmore, had been heavily damaged by the earthquake. New city ordinances required that the clinic designate a large percentage of its building space towards commercial purposes, a practice viewed as incompatible with its services. Clinic staff and directors interpreted the ordinance as an attempt to force the clinic out of its downtown location, thus reducing migrant worker visibility in the historic district. In a compromise, the clinic received approval from the Fillmore Planning Commission to rebuild the original clinic in the downtown area, complying with the city ordinance by having a retail pharmacy on its ground floor.

As part of its redevelopment, Fillmore also received a $1.2 million EDA grant to rebuild its damaged City Hall. Rather than replace the damaged structure, the city council approved a new design incorporating retail space for businesses displaced by the earthquake, as required by the grant. Although the building is now complete no retail businesses have actually rented space in it (Tyler 1997a). The use of federal funds and a central planning mechanism to rehabilitate the downtown, as essential as they may be in a small community, could not override the marginal economic status of some of Fillmore's small businesses. Numerous building owners could not afford to rebuild, even with the public funds available, due to the expenses involved in meeting current seismic standards and design code standards in the city. Thus as many as twenty-five vacant lots currently exist in the new downtown business district, where owners have razed buildings they could not afford to rebuild. At the same time prior Redevelopment Agency commitments to affordable housing have been put on hold due to problems in finding an appropriate location for the project.

Lacking both incorporated legal status and any local resources, Piru's reconstruction and redevelopment activities were heavily dependent on external organizations and funds. To provide organizational infrastructure for Piru's development efforts, the county established a county-wide Redevelopment Agency to manage development programs in the unincorporated areas of the county. In February 1995, county authorities met with Piru residents to establish a local subcommittee to liaise with the county Redevelopment Agency. While earthquake recovery was a priority, longer-range development plans, including expansion of affordable housing, were part of the agency's agenda. The County Redevelopment Agency targeted Piru with funds obtained from both the EDA and HUD. In 1996, using $200,000 grants from each agency, it began implementation of a comprehensive 'Piru Enhancement Plan' incorporating major infrastructural and business district improvements in the village (Community of Piru 1995; Ventura County 1996). The planning process, funded by the county, was managed by an architectural firm in collaboration with the residents of the community.

While initial funding of the Enhancement Plan was provided through block grants, long-term implementation would be funded by anticipated increases in property taxes expected to generate $4.6 million in new revenues over the next twenty years. The Redevelopment Agency used a needs assessment survey of residents to determine priorities in restoration and development actions as the community began the redevelopment process. Primary areas for attention included restoration of its small historic downtown, improvements in streets and pedestrian access, and home rehabilitation. As with Fillmore, Piru's downtown is being reshaped into an imagined historic railroad town. Responding to prevailing concerns about the safety of the Piru dam in earthquakes, the agency moved to acquire an evacuation staging area for future emergencies. For future economic development, the village will focus on the promotion of tourist attractions involving recreation at Lake Piru as well as in its restored historic core. In the short term, Piru remains tightly coupled to agricultural production in the area, and its attendant low wage employment.

The changes initiated in Piru and the funding to support those changes reflect the active support of Ventura County staff and the focused effort of one County Supervisor. A member of the Piru neighborhood council described the political changes in the village as a result of the supervisor's efforts:

> she was here all the time. She had all her county people out here for all the programs and, as I said, we had people who were really the top people who came out here . . . And all the other agencies stepped forward. Maybe some of them were around before, but after the

earthquake, they became more present here. Because of the earth-quake, we have all our county contacts; we know who to call now. They include us more; they remember us more when they're doing new programs. We go and represent our town, Piru, and see what we can get from them.

Prior to the earthquake, Piru was a poor rural village with few prospects for economic development, tied as it was to agriculture. The earthquake stimulated an influx of money and interest to the area, leading to what will be a significant reshaping and expansion of the community in the next few years.

Like Fillmore, Santa Clarita, had a Redevelopment Agency in existence prior to the earthquake. Santa Clarita's agency, originally authorized to begin renovations and infrastructural improvements in the older Newhall section of the city, had done little in its six years of existence. It lacked both a comprehensive redevelopment plan and a funding base. After Northridge, the city council voted to reactivate the agency – designating it a Recovery Agency – as a mechanism to promote earthquake reconstruction projects across the city. Although legally a redevelopment agency, the Recovery Agency designation reflexively linked it to earthquake-related projects and avoided the statutory limitations of the older redevelopment agency. The agency began directing its activities toward infrastructure improvements (primarily road building), housing and business repairs, and planning community revitalization in the older areas of the city. It initiated a low interest loan program for earthquake victims using a share of HUD block grants received by the city. The loan program was directed primarily at lower income households whose homes had suffered earthquake damage. The city estimated that as many as 5,000 households with sub-median incomes, particularly elders in mobile home parks, were in need of additional assistance beyond federal programs.

Unlike the Ventura County redevelopment agencies, the creation of Santa Clarita's Recovery Agency was immediately challenged by other agencies who opposed its claim on property tax revenues and CDBG funds. While reconstruction projects using HUD block grants, including Newhall revitalization, were not challenged, most long-range development plans using property taxes were. The key protagonist in the legal challenge was the Castaic Lake Water Agency, a regional water distributor, which filed a lawsuit against Santa Clarita's redevelopment plan. Water agency officials claimed that the redevelopment plans went beyond earthquake recovery and were part of a larger growth plan, and thus outside its 'recovery' mandate. More importantly, the water agency claimed that the Recovery Agency's use of property tax revenues would reduce Castaic's own tax revenues by an esti-mated $300 million over the next thirty years (*LAT*, 27 April 1994). Further opposition to Recovery Agency plans came from neighborhood groups that challenged a proposed expansion of highway infrastructure in Santa Clarita. (While the sprawling city has a meager infrastructure of highways and bridges to handle its growing traffic, homeowners' associations have aggressively resisted the construction of new highways near their neighborhoods.)

In response to this resistance, the Recovery Agency mounted a public outreach program to provide information about its proposed role in Santa Clarita's earthquake recovery and to solicit public opinion about the redevelopment projects. The overriding concern, however, was reaching an accommodation with the Castaic Lake Water Agency. In November, 1995, Santa Clarita and Castaic Lake Water Agency reached a tentative agreement in court. The city agreed to limit the Recovery Agency boundaries such that it would raise only $854 million in tax revenues, instead of the projected $1.1 billion (over thirty years) and agreed to reimburse the water agency for half of the estimated tax revenue it would forgo to the redevelopment projects. Irrespective of this compromise, the water agency filed a lawsuit to stop the Recovery Agency, alleging that the city was using the disaster as a front to embark on road construction and other development projects that had no connection to earthquake effects. In December 1995, the state appeals court halted Santa Clarita's earthquake recovery project on the legal basis that the city had not prepared a complete environmental impact report. In January 1996, the city council began a lengthy appeal process in the state courts to seek a reversal of the original court decision. Council members charged that the Castaic Lake Water Agency's ongoing legal challenges had retarded earthquake reconstruction in Santa Clarita by tying up the recovery agency's funds and projects.

In mid-1997 Santa Clarita lost its litigation with the Castaic Water Agency and had to disband its Recovery Agency, thus losing a development mechanism for long-term growth plans in the city. However, using its existing HUD funds, it established a new limited development program, one that did not threaten Castaic's expansive domain. Focusing development efforts entirely on the Newhall district, heavily damaged in the earthquake, in August 1997 it initiated its 'Old Town Newhall' development project. The goal is to renew older neighborhoods and business in the old urban core of Newhall, in a program akin to Fillmore's downtown redevelopment. While not specifically an earthquake recovery plan, it will nevertheless help revitalize the area heavily damaged in Northridge. As of November 1997, the project, much reduced from original Recovery Agency plans, had not met with noticeable resistance, either from Castaic or other political agents. While part of the development project will use tax revenues, the city anticipated community support for the new limited project.

The projects associated with the redevelopment agencies in all three communities address earthquake recovery while moving to establish broader development projects using public sources of funding. Santa Clarita's experience contrasts sharply with those of the Ventura County communities.

These contrasting histories point to emergent conflicts in initiating social and political change after disaster. In the case of Santa Clarita, the city used a state law designed to promote development in financially marginal areas and has proposed to use it both to assist in earthquake recovery and to promote economic development of a city whose annual median household income exceeds Los Angeles' by $20,000. By threatening the tax revenues of a locally powerful water agency the Recovery Agency became stymied by legal actions. While the Fillmore redevelopment plans proceeded with only minor legal challenges from historic preservation interests, they engendered conflict over the persistent issue of affordable housing.

Affordable housing and marginal households

While the city of Los Angeles promoted neighborhood retention and expansion of affordable housing primarily through the efforts of the LAHD, in Ventura County the drive for affordable housing was promoted by local NGOs and CBOs, drawing from locally available federal grants. As discussed in earlier chapters, the last thirty years have seen efforts by various organizations to promote low income housing in Ventura County, but these efforts have been marked by conflict and resistance pitting a dominant Anglo political elite against a largely poor politically disenfranchised Latino population. Support for low cost housing has come mainly from non-profit groups seeking to assist low income families, older retirees, and farmworkers. While several projects have been undertaken by CBOs in the Fillmore area to build low cost, cooperatively owned farmworker housing (Bee 1984; Menchaca, 1995; Peirce and Guskind, 1993), these efforts have consistently met with opposition from established political interests. This resistance to affordable housing has produced a spatial peripheralization of farmworker housing, which has been built in the interstitial spaces between communities and away from the Anglo middle class, a localized geography of exclusion (Sibley 1995).

Two county-wide CBOs that have been active in developing housing for the lower income households are Cabrillo Economic Development Corporation and Affordable Communities, Incorporated. Both used the earthquake and the subsequent resource availability to promote low income housing developments, albeit with mixed success. Cabrillo was originally established as a non-profit organization in 1968, committed to the construction of affordable housing for agricultural workers in Ventura County (Menchaca 1995). Cabrillo used political openings and resources generated by Northridge to propose a low income housing project in Fillmore. In May 1994, Cabrillo received a grant of $400,000 from the Fillmore's Redevelopment Agency as part of the $4.3 million in HUD block grants to the county. These monies were designated to fund Cabrillo's project for an 81-unit housing development just south of the downtown redevelopment district. The project was to include 30 apartment units designated specifically for farmworkers,

30 for very low income families, and 21 units of 'affordable' single-family homes for purchase. Priced at approximately $139,000, these homes would have cost nearly $100,000 less than the new home median for the region. Earthquake victims were to be given priority on both rentals and home purchases in the development. However, Cabrillo's project, although funded and approved by the Redevelopment Agency, met with resistance from the Fillmore city council. Council members and residents expressed concern about the location and potential 'social impacts' of the proposed development. After a series of negotiations, the landowner who had originally agreed to the sale of land for the development retracted the offer. At the time of writing, the project, although funded, had not been able to find a location that city planning authorities found suitable. As elsewhere in California, Fillmore residents, including landowners, continue to resist efforts to house agricultural workers in the very community that depends on wealth generated from their labor (Bolin and Stanford 1998). As the city reinvented its history with its renovated downtown, the presence of low income housing units nearby was deemed undesirable by development interests.

While Cabrillo's effectiveness was limited by continuing political opposition in Fillmore, Affordable Communities was able to negotiate its way past potential opponents, acquire land, and begin the construction of low income housing in Piru, which lacked Fillmore's entrenched Anglo power structure (e.g. Menchaca 1995). While Affordable Communities had their plans approved by the county prior to the earthquake, the disaster facilitated both funding and speed of implementation of the project. The original plan for 113 houses and thirty-five apartment units was developed in consultation with the residents of Piru, and it used their input to design culturally appropriate housing for the larger Latino households common in the area. Because of the housing crisis generated by the earthquake, the county streamlined the permit process, allowing construction to begin within a year of the main shock (subdivision permit processing can take up to ten years). In addition, the county gave Affordable Communities an interest-free $275,000 loan to pay permit fees. The company was also able to obtain an additional $500,000 through the county as part of a HUD HOME mortgage assistance program. Those funds will assist low income home buyers with down payments on the new homes when completed. Six of the thirty-five apartment units have been set aside for low income elders; these were purchased by the County Area Agency on Aging through a $360,000 grant from Ventura County block grants. Piru residents were given priority on rentals and purchases on all Affordable Communities housing in the village. By August 1995, the 35-unit apartment complex had been completed. Construction had begun on a new 33-home development, the first phase of the larger single-family housing project. In addition, Habitat for Humanity, which had actively assisted low income homeowners in Fillmore, Piru, and Simi Valley with home rehabilitation, secured $1.1 million in CDBGs through the county to fund the construction of an additional twenty-two new homes in Piru for very low income households. Interest-free mortgages will be made available to allow poor families to purchase the homes. Given these affordable housing projects, county officials projected that Piru's population would double within two years (by 1998) as a result. Thus, the earthquake promoted the substantial expansion of affordable housing in the rural village, making available both new homes for low income families, and the first affordable housing designated specifically for elders. As the new housing stock will meet current building codes, it should reduce residents' exposure to physical risks from earthquakes, and hence their vulnerability.

In Santa Clarita, as in Simi Valley, the bulk of the housing stock is homeowner occupied single-family dwellings. However in the older site of Newhall, there are large apartment and condominium complexes that house a mix of Latinos (including recent immigrants) and low income elders. The earthquake affected several multi-family complexes, with 673 housing units damaged and 87 units red tagged and vacated, marginally reducing Santa Clarita's affordable housing stock. While the city has moved to provide loans to assist apartment owners in rebuilding to preserve affordable housing, the process has been considerably slower than the restoration of Los Angeles' ghost towns, and one large apartment complex in Newhall remained abandoned and unrepaired two years after the main shock. Unlike Los Angeles, Santa Clarita has no rent control ordinance, creating concerns that rents for low income households would rise as apartments were repaired and owners attempted to recoup losses. In both Simi Valley and Santa Clarita mobile homes constituted the single major source of affordable housing. In Santa Clarita nearly 1,700 mobile homes were damaged or destroyed, affecting primarily lower income elders. The state run Mobile Home Minimal Repair Program was the main point of assistance in both suburban sites, and repairs were completed within a year for those who participated in the program. Neither of the larger sites has embarked on systematic programs to ensure the expansion of affordable housing stock. Cabrillo Economic Development Corporation did complete a 22-unit condominium complex affordable for lower income households in Simi Valley in 1996, marginally increasing the 'for purchase' housing stock available there.

While municipal authorities across the region are aware of the problems of affordable housing for growing low income populations, in the absence of a concerted federal effort to support such housing, most local authorities can do little to change the general historic trends. The earthquake, primarily through the efforts of HUD, temporarily provided new funds to initiate programs such as those developed by the LAHD and non-profit organizations in Ventura County. Although on an admittedly small scale, Piru experienced the largest proportional increase in affordable housing stock for its largely poor Latino residents. On the other hand, the Fillmore Redevelopment Agency's willingness to augment affordable housing stock has been unmatched by its city council and local economic elite. Both Simi Valley and Santa Clarita have used HUD funds to support restoration of damaged affordable housing stock, but neither has yet implemented longer-term projects to increase the availability of such housing.

Enhancing self-protection

While households and communities across the region were reconstructing homes and redeveloping business districts, efforts were also directed towards enhancing levels of social preparedness for future disasters. Such enhancement involved technically complex geological hazard mapping, development of new building code standards, and seismic retrofitting of older structures to meet new enhanced seismic codes (USGS 1996). Additional efforts in the research communities went towards improved emergency preparedness capabilities. Both Simi Valley and Santa Clarita upgraded emergency response capabilities with improved equipment and facilities, and both cities moved to more fully integrate the state Standardized Emergency Management System in their management procedures.

To improve self-protection measures among residents, various groups in the region initiated or expanded programs to increase household and community preparedness. In the study communities the disaster helped catalyze interest in developing volunteer disaster response teams. All three Ventura sites began to implement Disaster Assistance Response Teams (DART) that trained residents to prepare for and manage emergencies in their homes and neighborhoods. The focus of the training programs was to enhance self-protection skills and train citizen groups to assist in disaster until official emergency management personnel became available.

As part of the overall improvement of citizen preparedness in Ventura County, Fillmore and Piru were targeted for additional emergency training. The Ventura County Commission on Human Concerns (CHC), a community action agency, began offering education and training programs for the two smaller communities, after conducting needs assessment surveys to determine local concerns. The CHC, using a share of county block grants, presented earthquake preparedness programs as preparation for the formation of the DARTs. As a consequence of the spatial isolation of Piru and the specific concerns citizens had regarding the catastrophic flood potential of the Piru dam and lake, CHC offered additional training and equipment. Two critical elements were addressed: the provision of two-way radios to allow 'outside' communication should another disaster isolate the village, and the installation of a warning system, including sirens, that would be linked with the Piru dam. In interviewing Piru residents, the level of awareness of the St Francis dam failure of 1928 was notable and that collective memory promoted active concerns about the nearby dam. A Piru resident described an initial response team meeting:

Some of those [OES] people from LA were real impressed that we'd gotten to that stage of planning. I said, 'Well, if you had been in Piru at the time of the earthquake, and somebody calls you from the outside and asks you if your dam is breaking . . . What would you do? They got some rumors, they heard something. But, you are just sitting here in Piru and you don't know what's going on.' So, [with the warning system] we'll be able to find out what's happening. We'll be able to communicate with the dam operator up there. Right away, when something happens, we can call up and find out, 'Is the dam down? Is it leaking? Have you checked everything out?' Like that [aftershock] we just had on Tuesday night. I picked up the phone and called up to the lake. [The dam manager] told me, 'Yes, I've checked it out there. It's okay. You guys can go back to bed.'

To enhance social protection in the village, with CHC assistance, the team conducted a household survey of the community, identifying all residences, particularly elderly households, where people would need special assistance in the case of an evacuation.

In Piru, the lack of emergency management capabilities after the earthquake made community members particularly receptive to community-wide preparedness activities. A local activist commented on their situation and their new empowerment:

Since we're not incorporated, Piru belongs to the county. It's like they're responsible for us. Well, after the earthquake, they didn't know that we didn't have any electricity or gas . . . I felt that it was very important to start something in this community because I blamed it on us. We didn't know what to do right after the earthquake. It was our own fault. We should have had people like we do now who can go out and survey the damage . . . People who have contacts with the county . . . with emergency services, so they can tell the agencies that we're out of food, that we have injured people . . . So we've taken care of that now. We're responsible for communicating with [county] emergency services now.

Besides their county-wide preparedness training, the CHC also offered free installation of earthquake mitigation technologies in the homes of lower income residents in Fillmore and Piru. These free services included enhancements such as flexible utility connections, propane tank anchorage, water heater bracing, and securing of household objects. For households disadvantaged by very low incomes and for persons with physical disabilities, these hazard mitigation services were important new elements in reducing their exposure to physical risks.

Santa Clarita, through its OES sanctioned SECURE program, had a citizen preparedness training scheme in operation at the time of Northridge. The core of the plan was to train households to be self-sufficient for seventy-two hours following an emergency and to organize neighborhoods to assist their local areas until such time as official responders arrived. Some 5,500 residents of the city had been trained in the program and their response has been credited as contributing to the overall effectiveness of the emergency response in the Santa Clarita Valley. Through public outreach efforts, participation in the SECURE program doubled in the year following the earthquake. Project SECURE staff began further expansion of the program using outreach to identify and train households not reached in previous training efforts, particularly elders and Latino residents living in the Newhall area. SECURE organizers developed a simpler version of the program for elderly residents and produced instructional materials distributed to housebound community residents in retirement homes and mobile home parks.

NGOs and CBOs provided a second level of social protection in Santa Clarita by developing 'Caring Link', a volunteer network designed to coordinate and integrate disaster responsibilities among different civic groups in the Santa Clarita Valley. Organized by the volunteer Community Council of Santa Clarita, a coalition of more than thirty-five local civic groups, church-based service organizations, and social advocacy groups established service domains for future area disasters. Based on inventories of material and personnel resources, participating organizations divided and accepted responsibilities for assisting in the provision of emergency food, clothing, day care, and crisis counseling. These support services are coordinated with ARC disaster planning and are intended to enhance social protection in emergencies, particularly when the valley is cut off from LA County emergency services due to transportation route closures (SECURE 1992b).

ISSUES IN RECOVERY AND RESTRUCTURING

While there has been a tendency in US disaster research to treat recovery as a return to 'normalcy' (e.g. Tobin and Montz 1997: 183), evidence from both the Third and First Worlds suggests such views simplify complex processes (Maskrey 1994; Peacock et al. 1998). Analytical recognition of restructuring and recovery directs attention towards the social and political realities of community and state efforts to respond to a disaster's effects. As we have used it, 'recovery' designates those activities that attempt to address a disaster's immediate effects and reproduce the built environment and socioeconomic realms 'as they were' at the time of the disaster. Albala-Bertrand (1993) has referred these actions as a 'closed process' of containment and restoration of losses. Of course recovery actions can have unanticipated consequences that change the community in the process of restoring it. 'Restructuring' describes activities that represent structural changes in social and economic relationships, including departures from pre-existing social and political relationships engendered by the disruptions of the disaster. Such structural changes are understood as 'open processes', and include economic development (or deterioration), the reduction of vulnerability for marginalized groups, and other political-economic actions that alter future disaster losses (Albala-Bertrand 1993: 34). Disasters may constitute an opportunity to promote existing agenda and practices, they may exacerbate existing conditions and grievances leading to political mobilization, and they may disrupt power relationships, removing blockages to social action. We stress that the recovery/restructuring distinction is theoretical. No community is static, and both recovery and restructuring have roots in pre-disaster social dynamics and conditions. Disasters such as Northridge can have multiple and contradictory social, political, and economic effects, and it is difficult to allocate all issues unambiguously to either recovery or restructuring when they may involve components of both.

Recovery and restructuring take place in a political arena in which groups, both local and outside the community, negotiate and contest access to resources and post-disaster recovery and development agendas. While 'normal' processes of capitalist development and underdevelopment seldom rely on democratic inputs, the political disruptions of disaster may destabilize existing power relations and provide openings for popular inputs that challenge unregulated market forces. As diverse rebuilding tasks were planned and implemented after Northridge, a complex of federal, state, and local actors attempted to influence development agenda across municipalities. Evidence from Third World disasters shows that powerful political and economic actors, including dominant business interests and their political allies, may reimpose pre-existing political and economic agenda in attempts to reaffirm the status quo (Albala-Bertrand 1993; Anderson and Woodrow 1989).

However, it should not be assumed that the state's position is simply to promote recovery to the status quo after disaster. As seen in the case of the Ventura County government, local political authorities themselves may introduce or support social change agenda that benefit marginalized groups. In the county's case, new affordable housing that met current seismic codes was built, and financing was arranged to increase low income elders' and ethnic minorities' access to secure homes. The City of Los Angeles likewise actively pursued agendas to rehabilitate devastated neighborhoods, to increase the availability of affordable housing, and to increase the participation of non-profit housing corporations in the LA housing market (Comerio 1996; LAHD 1995; Stallings 1996). Northridge was also taken as an opportunity to introduce stricter seismic standards in building codes and actively move forward on overdue retrofitting of infrastructure, particularly highways, increasing levels of social protection across the region (SSC 1995; USGS 1996). At the state level, the government moved to provide a state-guaranteed insurance system that offers at least minimal coverage to citizens while insuring the continued presence and profitability of major insurance corporations.

The post-Northridge period eludes simple categorization by temporal or 'recovery phase'; each of the study sites is characterized by a complex and imbricated mix of local decline, recovery, and restructuring processes that are ongoing. Plans, programs, and local actions have been carried out by similarly complex assemblages of federal, state, and local government agencies, working in conjunction with the private sector, NGOs, CBOs, and community groups across the region. Since the complexity of these matters increases with scale, the processes are most apparent in the smaller communities. In both Piru and Fillmore, recovery and development were entwined as local authorities, activists, non-profit organizations developed, implemented, and sometimes contested programs and activities. Fillmore's business-focused redevelopment plan was in place before the disaster, and it was modified afterward to accommodate disaster effects. Yet the Fillmore Redevelopment Agency's support of a CBO's housing project was stymied by entrenched political interests. As it remade its business district, Fillmore invented a historic past, apparently excluding the poor working class that produced much of its real history. While NGOs that serve low income Latino workers contested this remaking and attempted to establish housing and health services in the 'new' downtown, their success was limited. On the other hand, federal funds, channeled through CBOs and NGOs helped poor residents repair their damaged homes and bring them up to current (and safer) building codes.

In Piru, sympathetic county administrators working with federal funds were able to promote rehabilitation of the village's small business district and speed the construction of already approved low income housing. Piru, which lacked the entrenched Anglo power structure of Fillmore, was able to develop and implement plans to enhance infrastructure and expand housing opportunities for vulnerable households without internal opposition. These developmental processes were made possible because of the new availability of funding as a result of the earthquake and the political mobilization of key organizations and political leaders. Of all the communities affected by Northridge, Piru has undergone the most substantial social changes. Although these changes may improve housing security and the amenities in the community, none fundamentally alter the social relations of production and associated economic inequalities. Similarly, the development of large low-income housing developments in Piru, with its majority of working-class Latinos and lower income elders, may simply reinforce existing patterns of 'ghetto-izing' poor ethnic groups away from middle-class Anglo neighborhoods of Fillmore.

Neither recovery nor restructuring processes necessarily address the socio-political roots of vulnerability but they do address more immediate risk factors. Since vulnerability derives from political, economic, and social processes that deny certain people and groups access or entitlements to incomes, housing, health care, political rights, and, in some cases, even food, then post-disaster rebuilding by itself will have limited effect on vulnerability (Dreze and Sen 1989). Rebuilding to higher standards that incorporate new hazard information will likely reduce the physical risks posed by hazards and thus possibly reduce aggregate disaster effects in future events. Yet people's access to safer homes and work places is determined by political and economic factors, not technological improvements or enhanced building codes. To the extent that the development plans incorporated into recovery processes improve people's access to employment/income over the long term, a reduction in vulnerability may be promoted, particularly when it is coupled with enhanced local preparedness. However, none of the recovery or development programs reviewed here is specifically intended to alter existing class relations or improve the economic conditions of most marginal households.

In the case of the larger sites, including the City of Los Angeles, households that obtained HUD housing certificates, specifically the 7,000 converted now to permanent housing subsidies, clearly had their access to housing improved. However, given the current neoliberal hegemony and efforts to abolish most forms of social welfare in the US, this housing entitlement is necessarily precarious, and not a substitute for more secure forms of housing access from improvements in income earning opportunities. Similarly, LAHD's efforts to convert an admittedly small portion of its affordable housing stock to the ownership of non-profit corporations improved housing availability for vulnerable households and may insulate them from predatory market forces. Efforts to restore businesses in Simi Valley and redevelop older sections of Santa Clarita could enhance income opportunities in the communities, although most new employment is in the lower wage service sector. Agencies and activists in Los Angeles and the four case study sites all recognize the persistent need for affordable housing, but improving access to housing requires political decisions and commitments that conflict with market principles. (The City of Los Angeles and Santa Monica's respective rent control ordinances are examples of attempts to keep housing rental costs affordable.) Northridge created circumstances and provided opportunities to address some of these issues, especially in the smaller communities. However, the larger historical and geographical trajectories of suburban expansion, immigration, economic marginalization, ethnic and class politics, anti-welfare politics, and ongoing economic restructuring appear to weigh more heavily on future patterns than the effects of a moderate earthquake.

NOTES

1 FEMA's Public Assistance Programs reimburse municipalities, counties, and eligible NGOs for repairs to public infrastructure and buildings, debris removal, and emergency protective measures. While FEMA will provide grants for mitigation to seismically upgrade buildings, they do not fund improvements such as enlarging buildings or other enhancements. Rather they will only pay to return the building to its pre-disaster condition. Further, cities must be able to pay for the repairs themselves, then bill FEMA, who will reimburse cities for allowable costs. Cities may or may not get reimbursement if they stray from FEMA guidelines and they have to assiduously document their expenditures. It is a time-consuming process and cities may wait a year or more before their reimbursements begin to trickle in. HUD and EDA programs, on the other hand provide the money 'up front' and allow considerable leeway in use of funds for development programs. Unlike the preceding, FEMA's mandate is quite specifically to assist communities in 'getting back to normal', not to fund community-wide enhancement plans.

2 These are totals to the Ventura County Board of Supervisors. They do not include grants issued to the Simi Valley City government, nor do they include assistance given by FEMA, HUD, and SBA to individuals in the county. Ventura County had earthquake insurance on county buildings in the county which reduced their need for public assistance grants for repairs.

3 Emergency managers and political leaders at all four peripheral sites made various claims about being forgotten or being ignored by FEMA. All believed that their respective communities were being overshadowed by the City of Los Angeles and its ability to command federal attention and resources. It is clear that Los Angeles was far better able to conduct preliminary damage assessments than other cities, and to present its needs to FEMA rapidly. Ventura County was not included in the initial declaration, having to wait two full days before federal aid was authorized, a point which resurfaced periodically in claims about 'being ignored' by FEMA. As with the City of Los Angeles, FEMA gave public assistance funds to Santa Clarita prior to receiving reimbursement claims. The $7 million given 'up front' to the city weakens its claim that it was low priority, as no Ventura County city received such funding prior to reimbursement claims.

7

VULNERABILITY, SUSTAINABILITY, AND SOCIAL CHANGE

Let us not, however, flatter ourselves overmuch on account of
our human victories over nature. For each such victory nature
takes its revenge on us.

Friedrich Engels (1959: 12)

The general approach we have explored is based on two related premises. The
first is that the root causes of disaster are found in the historical geography
and political economy of places. The second is that disasters occur at the
intersection of environmental hazards and vulnerable people, and as such are
social products. The hazards may be 'natural', technological, or some mix of
factors, but their risks and effects are mediated by prevailing social practices
and their material forms in a given place. Peoples' vulnerability to hazards is
produced by a combination of elements on different spatial and temporal
scales: these include underlying historical factors that shape patterns of social
inequalities and settlement patterns, and the more immediate and dynamic
aspects of politics, economics, environment, and culture in a place. Vulnerability to hazards is produced and reproduced through the changing nature of
people's access to resources and their exposure to risk (Wisner 1993).

As we have reviewed, in the one hundred and fifty years that the state of
California has existed, its landscapes and ecosystems have been radically
transformed by the forces of capital working in concert with state power
in a 'grandiose ecological project' (Harvey 1996: 185). The imbrication of
political and economic power, water imperialism, and nearly unchecked
urban growth have shaped and reshaped its social and physical landscapes in
an ongoing growth dynamic. The natural and technological hazards that work
their 'revenges' on California have become an increasingly prominent part of
those transformations. A century of real estate speculation supported by local
governments and public infrastructure has placed increasing numbers at risk
of earthquakes, floods, and fires, and those numbers in turn produce growing
pollution and waste disposal problems in the region (Davis 1992). It is then
left to public agencies to try to regulate the pollution and promote hazard

mitigation that is made necessary by virtually unrestrained private actions (FitzSimmons and Gottlieb 1996).

In the US, hazard mitigation and disaster response embraces a 'top-down' state-centered managerialism relying on scientific rationality and state regulatory powers, restricted by the imperatives of corporate capital and market forces. In California, which is far advanced compared to most US states, hazard management concerns itself primarily with technologically sophisticated emergency preparedness and technical/engineering solutions to weaknesses in built environments. It has had clear successes in promoting building and lifeline safety in the face of common disasters, although none of these systems has been tested in a large earthquake. Hazard management as a technically specialized field necessarily avoids the broader environmental and social contexts of disasters: their root causes in capitalist development, urban growth, and environmental degradation that place more people, their property, and livelihoods at risk of an increasingly complex array of natural and technological hazards.

The historical sources and current manifestations of vulnerability vary by place and hazard, yet there would appear to be consistencies in US examples regarding groups systematically disadvantaged and lacking response capacities. Studies of earthquakes suggest that the groups we characterized as persistently vulnerable are far from unusual – low income elders, poor Latino and African-American households, homeless persons, farmworkers, undocumented immigrants, and monolingual ethnic groups, have been variously identified as encountering problems in acquiring resources and recovering (e.g. Bolin 1994a; Bolton *et al.* 1992; Laird 1991; Morrow and Enarson 1996; Phillips 1993). Research on Hurricane Andrew identified numerous marginalized groups in Florida who suffered disproportionate losses and experienced exceptional difficulties in recovering. To varying degrees these involved disadvantaged persons akin to those identified previously; people whose already existing social and economic deficits produced barriers in recovery. In the Andrew research, these included lower income African-Americans, female-headed households, lower income elders, farmworkers, marginalized Latino ethnic groups, and Haitian immigrants – all of whom were made vulnerable by low wages, fragile homes and livelihoods, and limited access to post-disaster resources (Morrow and Peacock 1998). The existence of persistently vulnerable households raises issues of environmental justice and the unequal risks some bear as a result of poverty and discrimination (cf. Bullard 1990).

To develop a better understanding of vulnerability in other First World settings calls for new research which draws from a general framework for analysis such as the one presented in Chapter 2 (see also Blaikie *et al.* 1994; Cannon 1994; Wisner 1993). It requires an examination of disasters, not as extraordinary events caused by the intrusions of 'nature', but as socio-environmental processes historically prefigured and shaped both by patterns of social inequality and by the levels of social protection in a given place. Those at greatest risk to hazardous environments will be likely to include those who are disadvantaged by some combination of class, race/ethnicity, gender, and/or age (Blaikie *et al.* 1994). Their risks are greatest in the sense that they lack access to financial and material resources to cushion the impacts of a disaster and recover their losses. Without state or NGO interventions to augment their coping capacities, their lives may become increasingly precarious as resources are exhausted, homes are lost, and disadvantage mounts.

In Northridge, the majority of victims, of course, were not poor or marginally housed, and lived in areas with extensive emergency response capabilities and a reasonably secure built environment. For middle-class and wealthy homeowners, with secure incomes, insurance, and credit access, the risks faced from an event like Northridge will be far fewer than those whose lives are constrained by marginal livelihoods, inadequate entitlements, and insecure housing. While financially sound homeowners had their lives disrupted and incurred major expenses after Northridge, a significant minority had insurance, and most had access to commercial or federal loan programs to finance their recovery. Some reports suggest that many homeowners with earthquake insurance received overly generous settlements (Comerio *et al.* 1996; OES 1997), allowing not just reconstruction of properties, but enhancements beyond what their homes had before the temblor. The majority of federal assistance programs remain most accessible and most generous for 'conventional' middle-class homeowners (Bolin 1994b).

Northridge also gave indications that fractions of the middle classes are being made insecure as employment becomes increasingly precarious in the new regime of flexible accumulation – just as the industrial working class in Los Angeles and much of the US was 'deproletarianized' in the 1970s and 1980s (Anderson 1996; Davis 1992). Thus situational vulnerability may appear, the result of negative changes in the economic security of a household combined with hazard-related losses that cannot be managed with available resources. For a middle-class Californian, a house can be rapidly transformed from a major economic asset to an equally large liability by a combination of untoward economic and environmental events.

Vulnerability to disaster viewed across the complex social, political, and physical geographies of the world is obviously highly variable and requires attention to local conditions and their insertions within national and international political economies (Blaikie *et al.* 1994). We have drawn some broad comparisons between 'wealthy' states of the North and poorer states of the South, although this necessarily means ignoring the complex local geographies of risk. In aggregate terms, vulnerability in the First World is neither so deep nor persistent as it is in the Third World, there are more resources available, and levels of social protection are generally higher. Yet, as we have described, those resources are not necessarily accessible by the most vulnerable, and most resources made available after disasters do little to address the inequities and deficits that contribute to vulnerability and disaster to begin with. The dominant view informing relief programs in the US is that disasters are discrete and specific events with clearly delimited effects – and only those effects will be addressed by the assistance programs.[1] Federal relief agencies in the US go to great lengths to insure that grants are used only for losses and needs directly attributable to a specific disaster, thus decontextualizing both people and disasters. In no case will a victim of disaster knowingly be made better off than before the disaster using public resources (Collins 1997).

While our analysis has emphasized social factors in disaster processes, we are not suggesting that physical hazards in the built environments are unimportant. Environmentally triggered disasters occur precisely at the socio-spatial interface of human settlements and geophysical forces. The technocratic approach to hazard management first critiqued by Hewitt in 1983 still promotes a view of disaster as primarily a matter for science and engineering, focusing on understanding, predicting, and controlling physical effects through policy and regulation. In this physicalist view, disasters result from the vulnerability of structures, and not, as we have suggested, from the structures of vulnerability. The technocratic approach ignores the complex of social factors that make social protection unavailable to marginal households. As the catastrophic losses and disorganized emergency responses after the Kobe earthquake showed, excessive reliance on 'technological fixes' unevenly applied may produce a neglect of the human factors in social protection. In a world with rapidly expanding human populations, deepening income polarizations, growing poverty, and accelerating environmental degradation on a global scale, many capital- and technology-intensive 'solutions' to hazards are likely to only increase inequities in the availability of social protection, both within and between countries. 'Top-down' efforts to plan or engineer a safer human environment must confront the 'limits set by the economic and social inequities, cultural biases, and political injustices in all societies' (Blaikie *et al.* 1994: 219), for this is the realm in which human vulnerabilities are produced and disasters foreshadowed.

We have offered a preliminary sketch of the social elements that come together with a hazard to produce vulnerability and disaster in a specific urban region. The Northridge earthquake intersected with the changing social conditions in an area with some of the most dynamic political-economic and demographic restructuring in the First World (e.g. Soja 1996a). While Northridge was a totalizing experience for those households and neighborhoods damaged in its seismic convulsions, its aggregate economic effects taken regionally appear as a minor perturbation. This contrasts with devastating Third World disasters such as the 1976 Guatemala earthquake or the Mexico City earthquakes of 1985. In both cases, the disasters produced national crises with effects well beyond the immediate physical impacts (Albala-Bertrand 1993).

As the Kobe disaster has shown, a complex of factors can concatenate to produce large-scale losses in technologically advanced, wealthy nations, given a seismic force of sufficient magnitude located directly under a densely settled urban area. As in many Third World earthquakes, a major factor in Kobe's death toll was unsafe structures, in this case wood frame homes (Yamazaki *et al*. 1997). The high death toll of women and elders points to the intersection of cultural practices with unsafe buildings. Quite unlike poor countries of the South, the $150 billion in losses in Kobe were rapidly absorbed and state-funded reconstruction programs have made rapid progress in rehabilitation and reconstruction of buildings and infrastructure in two years. Of course the profound human costs of Kobe are not measurable on a metric of yen spent and buildings rebuilt, however impressive those have been. And in Kobe, those with the fewest resources, particularly elders and others on limited incomes, may well remain in peripherally located 'temporary' housing as they are priced out of their old neighborhoods by new, more expensive construction (IFRCRCS 1996). Kobe illustrated weaknesses in social protection, inadequate attention to hazardous residential structures and the fate of their occupants, and emergency preparedness capabilities inadequate to a large complex urban emergency (Hiroi 1997; Smith 1996). Each disaster offers lessons for those willing to consider them as they reveal weaknesses not only in the built environment but in the structures of the social realm.

DISASTERS AS OPPORTUNITY

While disasters involve the destruction of lives and property, in their erasure of the urban palimpsest something is created as well: opportunities to improve safety, enhance equity, and rebuild in new or different ways. Ideally, in our view, those opportunities would be used to produce safer communities with more equitable and sustainable livelihoods for people. The particular uses of urban space are always temporary, always changing, and reflective of power arrangements in societies (e.g. Harvey 1996). At its most general level, the entire process of modern capitalist development is one of 'creative destruction'; what came before – whether buildings, technologies, productive systems, urban spaces, people's livelihoods, businesses, and neighborhoods – is replaced piecemeal by that which is innovative, new, different, more profitable (Soja 1989). In their destruction, disasters produce change that inflects with ongoing social and spatial transformations and are best understood in the context of these processes. Perhaps nowhere is this more pronounced than in Southern California, a region which has grown in a century and a half from a few thousand Anglo-European colonists to a space-consuming urban region of 15 million.

The notion that disasters present opportunities for social action is well-known although its meaning is subject to a variety of interpretations (cf.

Smith 1996). They can provide opportunities to rebuild communities that are safer and that provide more livelihood opportunities. For the professional earthquake community, they are opportunities to push through more rigorous building codes and land use ordinances to make structure and lifelines physically safer against ground shaking (e.g. NRC 1994; SSC 1995). More negatively, disasters can also provide opportunities for politicians to enhance their reputations while doing little else; opportunities for organizations to acquire resources that are never delivered to the people who need them; opportunities for commercial exploitation by companies that raise insurance rates or landlords who raise rents; and, in the extreme, opportunities to intensify social inequalities and injustices (Blaikie *et al*. 1994; Hendrie 1997).

For disasters to be opportunities, whether to enhance social equity, environmental justice, and humanitarian values, or simply to promote intensified private accumulation, requires organization and resources. In the case studies, local organizations and federal resources came together in various combinations to provide assistance for those with unmet needs. In the City of Los Angeles, federal resources allowed the LAHD to target resources to multifamily housing to fund the rebuilding of low and moderate income housing stock and to prevent neighborhood deterioration and abandonment. Other cities, including Santa Monica and Santa Clarita were also able to use federal funds flexibly to address local housing needs through loans and grants for reconstruction. The small communities of Ventura County used federal resources to revitalize their destroyed business districts and develop low income housing opportunities. However, as seen in Fillmore, entrenched political interests may resist aspects of restructuring if they are viewed as a threat to their economic agenda.

From an international perspective, California might be viewed as exceptional in its disaster management capacities, its resource base, and its levels of social protection. It is a leader, even in the US, to the extent that it has substantial base of expertise and advocacy about disaster preparedness and management, and there are numerous organizations (government and NGO) in existence to apply that knowledge. More importantly, the state has been able, over much of its history, to appropriate federal resources for infrastructure, economic development, and disaster preparedness/mitigation. By most measures, Northridge was dealt with effectively, given its scale, and the short-term needs of vulnerable groups were at least recognized and resources made available by federal and state agencies. However, Northridge did reveal numerous unresolved issues including funding the reconstruction of multifamily housing, the disproportionate costs and liabilities associated with single-family dwellings exposed to moderate ground shaking, and weaknesses in lifelines and steel framed buildings. In view of more powerful earthquakes forecast for the region, these issues remain unresolved.

When the focus is broadened beyond the immediate temporal and spatial confines of the Northridge earthquake, there are fewer warrants for optimism: the state's persistently stagnant economy, its strident conservative political agenda that has severely constrained spending on public programs and services, a pronounced political hostility to immigrants, its growing environmental problems, and sharp economic and social polarizations are all prominent issues (e.g. Davis 1992; Scott and Soja 1996). As Anderson (1996: 358) writes, 'Extremes of poverty in Los Angeles rival the Third World . . . ' noting that the gap between rich and poor in LA approximates that of Mexico City. Against that economic background, funding for hazard mitigation and earthquake safety programs is often in short supply, particularly from the view of state risk managers (SSC 1995). Similarly, if the federal government had not covered half or more of the incurred losses of recent California disasters, recovery after disasters like Loma Prieta and Northridge would be far slower and more inequitable than they are proving to be, irrespective of the role of private insurance.

In the political and social flux that can follow a large disaster, the potential for social change is immanent, but as the literature suggests, too often the impulse of governments and citizens alike is to return as quickly as possible to what was, particularly among those who gained privilege from the previous arrangements (Albala-Bertrand 1993). If vulnerability is to be addressed, it requires directing attention to those fractions of the population with the least privilege, those inhabiting unsafe structures and dangerous areas, and above all, those with the most fragile livelihoods and lowest incomes. To make opportunities to reduce vulnerability will mean linking disaster recovery, in the limited sense of repairing damages, with programs designed to mitigate risks and expand economic opportunities for those with the fewest resources (e.g. Maskrey 1989).

Disasters present opportunities to build safer communities by reconstructing structures in ways and in places more suited to the particular kinds of hazards they may be exposed to. In earthquakes, this involves building more seismically resistant structures that will protect inhabitants from death or injury in the event of earthquakes. In the reconstruction process after Northridge, damaged structures have been required to meet current building codes, and damaged mobile homes braced for security against future earthquakes. This somewhat optimistically assumes that all builders fully conform to new codes and standards and that all owners have been able, or will be able, to fully repair their homes. Building codes have been enhanced and retrofit ordinances have been passed in various municipalities throughout the region in response to the structural damage. In our study sites, low income homeowners could not, without special assistance, afford repairs, much less seismic upgrading to current codes on their homes – in such matters safety does have a price.

The knowledge of how to seismically upgrade is well established and widely disseminated in California. However because implementation is potentially expensive, raising construction costs and rents, applying the knowledge for safer structures is uneven at best. Cost-cutting construction and shoddy building techniques occur in spite of building codes and hazard mitigation knowledge, as both Northridge and Hurricane Andrew have recently demonstrated (SSC 1995; Peacock *et al*. 1998). Left to itself the market is unreliable and discriminatory in matters of social protection. In the First World, as in the Third, it is those with the fewest resources who can least take advantage of safer building technologies, unless extensive efforts are made to direct special assistance to the most vulnerable.

The key issue is how to build safer communities that are affordable for all who live there, not just the privileged classes. This, we suggest, requires both a sensitivity to local sociocultural complexity and a *participatory approach* to planning and mitigation (e.g. Maskrey 1994). Participation can create the illusion of democracy if it is not among equals and if it is used merely to ratify elite or technocratic agendas (Milton 1996; Szasz 1994; Woods 1995). Disasters such as Northridge become focusing events for both policy making and hazard mitigation activities (Birkland 1996) and provide openings to instigate new building regulations and land use practices. The risk in such matters is when the disaster is separated out from the ongoing flow of daily life and treated as a thing in itself, the everyday needs of people will be ignored. The excessive privileging of scientific and technological discourses and bureaucratic complexity in hazard mitigation and disaster recovery creates barriers to democratic participation and local empowerment. As described in the study communities, CBOs and NGOs became important mediating structures between the technical and bureaucratic apparatuses and people at the local level.

In the US and other wealthy countries with a great deal of scientific and administrative specialization, it would appear rather too easy to mistake one's particular professional technocratic agenda regarding disaster reduction and hazard mitigation for one that is universally shared and of primary importance. Often a tone is adopted in reports on hazard mitigation in the literature that implies an ignorance or wilfulness on the part of people when they don't voluntarily implement hazard reduction strategies or acquire earthquake insurance. Thus Comerio *et al*. (1996: xvi) refer to California hazard mitigation efforts as having 'fallen on deaf ears' of the public. Indeed, in criticizing the expanded federal presence in disasters, they write that the 'federal attempt to be all things to all disaster victims was leading the US to become a nation of whiners and con artists, out to fleece the federal government . . . as if the hurricane/earthquake/storm flood were somebody's fault' (Comerio *et al*. 1996: 62). This curiously hostile stance to disaster victims criticizes federal disaster assistance for having been 'transformed into entitlement programs for aggressive individuals and [local] governments' (Comerio *et al*. 1996: 62). What these criticisms ignore is that ultimately federal interventions are designed to restore the conditions of capitalist accumulation and economic functioning, and assisting victims is one component of that larger and more fundamental project.

Blaming victims under the neoliberal 'personal responsibility' catchphrase is growing in popularity in the US, even apparently among some in the research community. Given that the majority of those displaced in Los Angeles County were apartment dwellers, there would appear little that they could have done to insure that their apartment building incorporated the latest seismic mitigation technologies. As the case studies indicated, people, particularly low income minorities, elders, and financially stressed middle-class households, often have to balance limited resources against the constant hazards of illness, unemployment, rising living costs, and even homelessness. Given persistent general risks, it should not be overly surprising that many appear willing to 'take their chances' against the infrequent earthquake or occasional flood and do not purchase expensive but minimal coverage insurance. As Hewitt asked (1983b: 26): 'Do [people] appear "poorly adapted" to us because the socially narrowed world of technocratic or academic specialists leave us incapable of recognizing the realities with which other persons and groups must deal?' If disasters are to be taken as opportunities to make safer communities, developing an informed understanding of the local realities of daily life surely must be as important as imposing new building codes, zoning regulations, or insurance requirements. Efforts to discipline disaster victims by denying them assistance if they have not taken adequate self-protection measures may fit neoliberal ideological agendas, but are likely only to increase the social inequalities that are too often already amplified by disaster.[2]

Disasters present opportunities for mitigation and risk reduction by temporarily increasing the salience of environmental hazards while also offering opportunities for implementation while people repair and rebuild. In the First World, persistent issues include affordability and perceived effectiveness of various mitigation strategies. As Palm (1995) notes, earthquake insurance tends to be expensive, even for middle-class homeowners, and that combined with large deductibles on claims sharply reduces its appeal. The regional purchase of earthquake insurance in California has gone up after recent earthquakes (Bolin 1994b; Palm 1995), which, in some measure reduces the financial risks faced by homeowners. However the earthquake insurance offered by the California Earthquake Authority is marginal at best and has little to offer low income owners and the increasing numbers in California who can no longer afford to purchase a home.

RESOURCE ACCESS ISSUES IN CONTEXT

We have described the large and elaborate relief and recovery apparatus which has been available in recent large US disasters. It is a system that has improved in recent years in its delivery of short-term assistance to an increasing socio-economically and culturally diverse population in the US. It is also a system that is being shaped to conform to changing ideological regimes. We reviewed evidence of this in the new exclusionary clauses in federal assistance regarding undocumented immigrants and others lacking admission qualifications. As our case studies as well as other recent research have demonstrated (Morrow and Peacock 1998), it is a relief system that is slowly adjusting to the diversity of household organization now common in the US. The fact that poor households may have more than one 'head of household', that they may have extremely limited documentation about their residency, or they may occupy structures that FEMA doesn't recognize as habitable, all have created problems for low income and ethnic households in acquiring even minimal federal assistance. The administrative requirements of federal bureaucracies are often at odds with the realities of day-to-day life for those who have been marginalized by poverty and discrimination.

After Northridge, numerous NGOs and CBOs were able to offer varying degrees of assistance to vulnerable households in ways that were responsive to people's diverse capabilities and needs. There were some voluntary organizations that relied solely on their own resources and the efforts of their volunteers to provide needed services to victims. By not using federal resources, they were free to assist marginalized individuals, particularly undocumented immigrants, in ways that did not necessarily conform with federal rules and procedures. Other community-based organizations that contracted with federal agencies, such as those involved in HUD's Mobility and Mobility Plus programs, were able to use their knowledgeable local staffs effectively to work with vulnerable households in community settings where they were already legitimately established. Federal flexibility in funding and in networking with the non-profit sector enhance the delivery of services over what was available solely through the DACs.

On these issues, HUD, through its relaxation of programmatic requirements on the disposition of Community Development Block Grants, allowed local governments and redevelopment authorities to fund special recovery programs using CBOs as the service providers. However, under a bill with the Orwellian title 'Housing Opportunities and Empowerment Act', the dismantling of HUD has recently been proposed (GAO 1997). The conservative attack on HUD would eliminate many of its programs for low income households and drastically cut funding on surviving programs given over to other agencies. Many federally supported housing projects for low income residents would likely become fiscally unviable and the occupants would be forced into the more expensive commercial housing market. It would eliminate a critical resource for assisting communities with low income housing, and would also constrain availability of federal resources for middle income homeowners through federal mortgage guarantees (GAO 1997). Any reduction in the availability of rent supplement programs and affordable housing projects has the potential of intensifying the 'affordability crisis' prevalent in many US cities (e.g. Wolch 1995).

The disaster-specific focus of federal programs and technical specialization of hazards managers generates something of a collective tunnel vision that ignores the larger set of social and environmental circumstances that produce risks and complicate disaster response. Federal efforts to support structural mitigation activities work to reduce physical risks and losses, but these are programs that lower income populations are unlikely to benefit from. More general issues of insecure and low wage employment, lack of affordable and safe housing, weak local economies, and the other factors that create insecure livelihoods are outside the purview of disaster management, however much they may contribute to the shape that disasters take. To the extent that such conditions are exacerbated by regressive welfare policies in the US and elsewhere, vulnerabilities may well increase as federal entitlements disappear (e.g. Demko and Jackson 1995; Wolch 1996). These trends create increasing uncertainties as to the demands that may be placed on community-based organizations that will serve marginal households in future US disasters (Wallrich 1996).

Rather than confining them within increasingly narrow and specialized approaches, hazards and disasters should be seen in the larger historical and ecological contexts. In basic principle, this means the coupling of disaster mitigation with development plans informed by attention to questions of environmental sustainability (e.g. Blaikie *et al*. 1994; Mileti 1997; Peacock *et al*. 1998). Community-based organizations and NGOs, by working in civil society, can empower people to participate in reducing the risks they face and promote equitable development plans. Empowerment of local citizens can be an important part of reducing future vulnerability (Maskrey 1994) however ideological or unscientific it may sound to those who embrace a state-centered technocratic approach (Hewitt 1997). Leadership training and grassroots organizing can be an important component of reducing vulnerabilities by increasing choices people have in their daily lives and in controlling the risks they are exposed to.

FISCAL CRISES AND LONG-TERM PROSPECTS

Perhaps symptomatic of the postmodern condition, the dismantling of welfare programs that help house, feed, and provide health care for low income households, elders, and children in the US is proceeding apace, part of the effort to 'balance' the federal budget. The restructuring of tax codes to further advantage the very wealthy and continued very high levels of defense spending, suggests that the budget is being balanced at the expense of those who can least afford to pay (Abramovitz 1996; Ehrenreich 1997). Apart from humanitarian concerns such a regressive political dynamic portends, from a disaster vulnerability perspective, it suggests that these federal actions could increase the number of 'at-risk' households in the US (Demko and Jackson 1995) and create newly vulnerable households. The so-called 'safety net' to prevent absolute destitution in the US has been effectively dismantled with the passage of the 'Welfare Reform Act' of 1996, and welfare responsibilities have been shifted downward to individual states. Access to public assistance is now strictly limited and a lifetime total of sixty months of assistance is allowed. New criteria of eligibility do not consider need but rather age, employment status, previous assistance received, immigration status, and personal behavior (Collins 1997). As a result the demands on NGOs in assisting the very needy are likely to increase significantly in the US. In this realm, disasters clearly are a minor issue compared to the larger concerns for the welfare of these newly 'disentitled' households.

This entitlement elimination spills over into disaster entitlements as well and recent changes indicate a scaled down and increasingly restrictive federal assistance package in US disasters may eventuate. Congressional debates over Northridge funding and more recent disputes over budget allocations for severe flooding in North Dakota in 1997 indicate that humanitarian assistance in disasters is no longer exempt from the conservative ideological agenda. The greatest effects of a scaled-down federal involvement likely will be precisely in the area where federal involvement needs its greatest improvement: long-term recovery. Moves toward a federally funded or federally guaranteed hazard insurance are unlikely to reduce federal disaster obligations (US Senate 1995), and would offer financial protection only to conforming property owners. However these are concerns that have not as yet affected major federal disaster response operations, rather only signs of future restrictions.

In the case of earthquakes, the reliance on the 'free market', through insurance and commercial loans as a substitute for federal assistance seems a questionable solution likely to create further disadvantage to marginal households. The market failures in providing secure, affordable housing and affordable earthquake insurance in California are well known. That private insurers paid out $12 billion in claims should be balanced against their cessation of the sale of earthquake insurance after Northridge, and the thousands of disputed claims that finally prompted the creation of a state mediation authority two years after the main shock (CDI 1997). The market does not serve those who cannot pay and private companies will avoid insuring households unless risk exposure is minimized and profits assured.

While privatization may be the favored neoliberal solution to most social ills, market logic appears antithetical to humanitarian concerns for the welfare of disaster victims. People's vulnerabilities are reduced by increasing the kinds of choices they have in their lives – choices in employment, education, housing, health care, family planning, safe environments – not by foreclosing choices by increasing costs and cutting entitlements (UNDP 1990). Markets provide consumer choices only for those already privileged by market transactions; for vulnerable households, those who have been marginalized by economic and cultural practices, collective actions are needed to protect and secure fragile livelihoods and the environments upon which they depend. To address underlying causes of vulnerability will take ethical and political commitments well beyond market imperatives or the narrow focus of disaster mitigation (Blaikie et al. 1994).

DEVELOPMENT, VULNERABILITY, AND SUSTAINABILITY

The case studies illustrated some of the possibilities and limits of development and redevelopment in the context of a disaster. Our historical sketch of California's dynamic expansion illustrates that development and economic restructuring are continuous and the effects of any disaster are a (small) part of those dynamics. Across the Los Angeles region, the earthquake became a mechanism for resource acquisition to promote various programs of reconstruction and development. In some instances, redevelopment agencies targeted the enhancement of business districts, while a number of cities, particularly Los Angeles, focused on insuring reconstruction of multi-family and low income housing. Outside of the City of Los Angeles, recovery and development plans have taken as key goals population growth and increased commercial activity. As outlined in Chapter 3, growth and unrestricted accumulation have been the fundamental historical trajectory in the state, and hazardous environments have done nothing to deflect that (e.g. FitzSimmons and Gottlieb 1996). Nowhere in the thousands of pages of documents, reports, and media sources reviewed in the course of our research has there been a suggestion that California's disasters are evidence of the limits of growth in the state or that there are enough people at risk of environmental hazards. Indeed much of the technical engineering, geological research, and seismic safety advocacy, because they assume risk is a function of hazardous structures alone, suggest that technical fixes along with land use planning will permit 'safe' future growth. Yet too often growth in the late twentieth century is coupled with growing social inequalities, accelerated environmental degradation, and increasing vulnerability of marginalized groups to environmental and economic crises (e.g. Escobar 1996; Peet and Watts 1996; Soja 1996b).

Discourses of environmental justice and sustainability have gained considerable currency in environmental writings over the past decade (Bullard 1994; Harvey 1996) and are filtering into the hazards and disaster literature as well (Blaikie et al. 1994; Mileti 1997; Oliver-Smith 1996). As a normative position, environmental justice argues that environmental safety and health cannot be a luxury reserved for the privileged classes or wealthy countries. On our reading, questions of environmental justice raised in the spheres of technological hazards and resource development projects (e.g. Gottlieb 1993; Johnston 1994; Sachs 1996) are equally applicable to natural hazards. Marginalized people unevenly bear the costs of disaster just as they are more likely to be exposed to technological hazards and to suffer the effects of resource development schemes (Bullard 1990; Hecht and Cockburn 1990; Shiva 1989). And while such risks are unevenly distributed between North and South, they are also inequitably distributed within countries. Environmental justice, in this sense, must be understood as a universal human and community right to a 'secure, healthy, ecologically sound environment' (Aguilar and Popovic 1994: 19). Framed as such, environmental justice principles concern exposure to risks of the natural environment just as they do exposure to environmental toxins and to destructive development projects.

Questions of environmental justice connect to the current discussions of sustainable development. While sustainability is subject to wide and contradictory interpretations, there is general agreement that it embraces social and economic practices that can be carried out indefinitely into the future without shifting environmental and social costs on to other people, ecosystems, or generations (e.g. Pirages 1994; WCED 1987; Yanarella and Levine 1992). Disasters reveal both patterns of social vulnerability and ways that economic activities and their material expressions have created environmental risks. As a physical occurrence, disasters provide an arena, given adequate organizational and fiscal resources, to engage various sustainability principles in reconstruction and development (Mileti 1997; Oliver-Smith 1996). Yet, as our case studies showed, most reconstruction is market driven, and what planning there is focuses on promoting growth and facilitating enhanced accumulation. Disasters and other environmental problems are too often treated, not as symptoms of more basic political and economic processes, but rather as accidents whose effects can be remedied by sufficient application of technical skill and knowledge. The discourses of expert knowledge in turn marginalize and disadvantage those with the least command of it. While it is the language of business, government, and academics, it is not the language of those most often at risk of environmental hazards (Harvey 1996).

While growth is ineluctably linked to capitalist ideology, sustainable *development* should be disarticulated from the notion of sustainable *growth* if the former is to have any distinctive meaning in the context of the industrialized states of the North (Daly and Cobb 1989). For Daly and Cobb, sustainable development involves *qualitative* changes in economic systems, not quantitative expansion of the system. Yet the potentially vast financial risks that California or Japan face from earthquakes are precisely a consequence of the continuing quantitative expansion. It is recognized that much of the economic growth in the North has come both at the expense of the South and of the global environment in general (J. O'Connor 1994; Peet and

Watts 1996). The North, in particular the US with 4 per cent of the world's population, consumes vastly disproportionate shares of non-renewable resources and is the major producer of greenhouse gases and ozone-depleting chemicals (Sachs 1996).

Vulnerability is one consequence of these unsustainable economic processes. As Blaikie and his colleagues (1994: 233) write, 'vulnerability is deeply rooted, and any fundamental solutions involve political change, radical reform of the international economic system and the development of public policy to protect rather than exploit people and nature.' Broadly speaking, to persist in unsustainable practices that degrade physical environments is to increase risks to inhabitants of those environments over time. These risks range from increased severity of hazards because natural systems are destroyed, to economic crisis due to exhaustion of non-renewable and mismanaged renewable resources. Another outcome in this process is the decline in environmental quality, often in the form of air and water pollution, in the course of growth.

In California and much of the western US, growth is space consuming, relying on vast networks of highways, private automobiles, and heavy dependence on non-renewable resources to fuel the system (Bottles 1987). In this process both natural areas and agricultural lands are 'converted' to suburban development and more people, structures, and lifelines are spread out across the fault zones and fragile ecological areas of the arid west (Davis 1996; Gottlieb 1993). Disasters such as Northridge provide opportunities to rethink such growth trajectories and to explore questions of sustainability and the limits to growth, particularly in cases where redevelopment projects are undertaken. Yet these are opportunities not taken because even in a state like California, there is only limited pre-disaster planning for recovery, leaving most communities in a scramble to devise recovery strategies in an *ad hoc* manner after disaster (Bolin 1994b; cf. Tyler 1997b). In such circumstances, questions of sustainability will be ignored, as they are in most areas of planning and development in the US. Any suggestions of the limits to growth or sustainable development that might require such limits go against powerful economic interests that profit from consumption and the waste that it produces (e.g. Horton 1995; J. O'Connor 1994).

REDUCING VULNERABILITY

A series of disaster case studies on one unique and very large urban area in a wealthy Northern country does not provide, by itself, a comprehensive assessment of the utility of vulnerability analysis. Vulnerability analysis is an analytical approach rather than a grand theory of disaster: it asks us to examine conditions that place people at risk, conditions that are themselves grounded in the history and geography of places and that are the ongoing results of human practices. While cognizant of the limits of the research, we have reviewed evidence from both Third and First World studies that support the validity of a vulnerability approach. In the end, the key concern is less with what 'works' for disaster researchers than with what works to guide practice in the search for more secure lives for people. Thus we urge a *precautionary principle*: enhancing social protection and reducing physical risks for vulnerable people while working to address the root causes of their vulnerability should be part of all disaster preparedness and mitigation, North or South. Vulnerable populations, in their localized diversities, require focused attention of community organizers, development specialists, hazard managers, and governments, if a sustainable reduction in vulnerability is to be achieved (Blaikie *et al.* 1994). One focus should be on empowering disadvantaged groups to enhance the choices they can make in their lives and to relieve the constraints of poverty, discrimination, and political marginalization. Technical expertise should be cooperatively combined with local knowledge and an understanding of community needs to produce safer environments.

Since the forces that produce vulnerability are ongoing and dynamic, situations are fluid and planning must be flexible. As discussed in Chapter 2, a weakness of the conventional view of disasters was the assumption that communities had a normal and stable state prior to disaster. A sudden increase in unemployment in one economic sector, a change in immigration patterns, or a shift in political regimes can very quickly alter the risk profile of specific groups and communities, however normal such things might be. 'Normalcy' is too often conflated with natural or desirable or 'functional', and taken as something to return to after a disaster. A normal state of affairs may be an ongoing disaster for poor and disenfranchised peoples subject to the whims of the international political economy (Escobar 1996; Maskrey 1989). Risk and disaster then are a part of a web of insecurity that some find themselves in – people who have already become victims of 'economic disadvantage, uprooting, chaotic urbanisation and political change' (Hewitt 1997: 359).

For those who have gained cultural capital through their technical expertise in increasingly narrow aspects of disaster, the broad agenda of vulnerability studies, its concerns with social justice, local knowledge, and local capabilities, may well be judged too radical or utopian. Yet those who claim technoscientific privilege above the capabilities of ordinary people run the risk of 'perpetuating a paternalistic and colonial mentality' (Hewitt 1997: 353).Technical expertise left to itself is exclusionary, lending itself to centralized organizations and systems of dominating knowledge (e.g. Foucault 1980). As evidence assembled here indicates, after Northridge the needs of vulnerable people were often best served by community-based programs that attempted to remove the barriers of expert and bureaucratic systems, and provided resources that fit local requirements and capabilities. It was at this intermediate level of organizations in civil society that state-based resources and technical knowledge were applied flexibly and in response to local conditions of vulnerability. Yet, without the availability of state resources or a large donor pool, NGOs and CBOs will be limited in the extent and length of support they can provide to vulnerable households.

Governments, through their abilities to socialize the costs of disasters, are an obvious starting point for programs of vulnerability reduction and hazard mitigation. Yet they may lack the political will, resources, or ethical commitment to do so. While the US government, through the particular efforts of FEMA, has taken an activist role in disaster relief and management, it has rapidly withdrawn from other areas of social welfare and housing support. There appears a growing incoherency between the intentional disadvantaging of US welfare recipients on the one hand, and the still high levels of assistance available to disaster victims on the other. Increasingly the only 'deserving poor' worthy of support are those caught up in disaster, although the assistance received will do little about their cumulative social disadvantage.

A reduction in vulnerability must be seen to go beyond reducing the physical risks people face from hazards. Modifying the built environment to be more secure and doing so in an accessible and equitable fashion is an important component of hazard mitigation. Government agencies at different levels may pass innumerable laws, codes, and standards, ostensibly to make buildings more secure. However, as Blaikie *et al.* (1994) note, the key is to implement and manage the rules and standards to produce systematic reductions in people's exposure to environmental risks. Regulations without implementation and enforcement may do little good, however sophisticated the regulations are. Similarly, if risk reduction is affordable or available only to those with certain levels of income, then disadvantage may be increased for marginal households. Political commitments, whether through governments or NGOs, should be developed to direct mitigation efforts to those segments of the population whose lives are most precarious and endangered.

More generally, reductions in vulnerability will come from improving the access of marginal households to resources, to promote more secure lives and livelihoods. Improvements in education and income opportunities will allow such households to also reduce physical risks they face by increasing their choices over where and in what they reside. In terms of the progression of vulnerability (Figure 2.1), these are matters that concern economic and political power in societies and the legal rights and social protections available to people. Reversing the effects of long histories of ethnic discrimination and economic marginalization requires major reforms unlikely to be implemented under current political conditions in the US. However disasters do, on occasion, provide political openings to include marginalized groups and promote new forms of development that increase access to income earning opportunities for previously excluded households. The actions of sympathetic local government agencies and activist CBOs/NGOs can be important in promoting local programs that enhance the security of marginalized households. Vulnerability reduction will involve changing economic relations through environmentally sound and locally accessible development programs. The importance of grassroots citizens groups and locally responsive NGOs has been well-documented in movements to preserve local environments, protect livelihoods, and combat technological hazards (Bullard 1994; Shiva 1989). The same forces in civil society that have mobilized against toxic wastes and for the preservation of healthy local environments (Szasz 1994), have significant potential for organizing to protect their communities against natural hazards in 'community-based mitigation' programs (Maskrey 1989). Since popular and political interest in disaster mitigation rapidly wane as the emergency winds down, making hazard reduction and social protection activities an integral part of a larger agenda of environmental protection and economic development is a strategy that ensures continuing attention.

From a vulnerability perspective the major implications of Northridge involved housing and the lack of institutionalized mechanisms in the US to rapidly and equitably fund the restoration and enhancement of housing stock. For homeowners with insurance coverage and resources or credit to pay deductibles, the major issues were the speed and adequacy of their settlements, and finding a contractor to repair their homes. While not downplaying the many struggles such households may have faced in restoring their homes, this is generally that segment of the US population that remains best served by government programs, commercial credit institutions, and insurance services. The greatest problems were in financing and repairing apartments, particularly those with marginal cash flows (Comerio 1996). Without the special funding arrangements made possible by HUD, the Los Angeles housing crisis would have been significant, with a critical reduction in affordable housing. To protect this housing sector, funding mechanisms should be in place to rapidly implement the repair and upgrading of structures that offer affordable housing to low income groups. Commercial lenders and private insurance are not affordable or available for building owners with marginal incomes. Thus funding for this housing sector should be quickly available in the form of grants and very low interest loans from the federal and state governments. Having to assemble *ad hoc* programs after the disaster is time consuming and outcomes are uncertain. It also requires specialized skills and technical systems not necessarily available outside of sophisticated urban governments such as that of Los Angeles. As the occupants of lower cost housing will typically have the fewest options for alternative housing, insuring the continued or expanded available of such housing is an important element of social protection.

While the restoration of single-family dwellings is normally more rapidly accomplished than that of multi-family structures, lower income homeowners and economically distressed middle-class owners faced significant constraints. For those ineligible for either commercial or SBA loans, sources of reconstruction funds were limited, constraining the options people had for recovery. For these economically constrained households, an enhanced system of housing grants, federally funded but distributed locally by housing authorities and NGOs, in amounts large enough to repair lower cost housing is warranted. For marginal households, that can little afford additional monthly payments to banks, combined home repair and mitigation grants of adequate size would encourage replacement of housing stock built to more robust standards. While CBOs and NGOs provided a critical intermediate level of recovery support for lower income households, they generally lack consistent or predictable funding. An expanded government willingness to fund programs in this sector could improve the efficiency and effectiveness of federal spending, by allowing locally knowledgeable organizations to provide appropriate assistance. While a more generous federal low income grant program would not address the causes of vulnerability, it would work to prevent deterioration in the living conditions of marginal households while allowing homes and apartments to be safer in the face of future disasters. This is an area where the state should have a fundamental moral commitment to insuring that people can be safely housed at prices they can afford.

We recognize that what is politically feasible or affordable to reduce environmental risks in a wealthy state is very different from what may be possible in poorer countries of the world. Similarly a suspicion of the power and motives of modern states is necessary: while they can do much to improve the lives of people, the potential coexists for them to do unspeakable harm (Foucault 1980; Hewitt 1997) The problems of Northridge and the mechanisms developed to deal with those problems are not necessarily transferrable to other hazards or other societies. While the political-economic forces that produce vulnerability are at work everywhere, vulnerable groups will necessarily vary by social, cultural, and geographic contexts, as do the hazards they may be exposed to. Thus there are no grand lessons to be learned from studying the experiences of a diversity of people in several communities in one US state. The various factors that serve as markers of vulnerability may be generalized, but the particular ways they combine require an attention to local complexity and diversity. To reduce vulnerability requires expanded understanding of the ways societies unevenly allocate the environmental risks and the social and political commitments to promote greater economic equity and environmental justice.

NOTES

1 FEMA has recently increased its emphasis and support for hazard mitigation activities in communities, to reduce disaster severity in future calamities. It has also begun a pilot program of 'disaster resilient communities', offering grants to selected municipalities to assist in hazard mitigation and emergency preparedness

activities on a community-wide basis. This growing emphasis on incorporating hazard resilience in a more holistic fashion, coupling it with general planning and development processes is an important new direction in the US.

2 FEMA recently announced a change in the National Flood Insurance Program policies. FEMA will now provide one-time coverage to uninsured property in a designated flood plain. The property is permanently banned from future individual assistance if insurance is not obtained. The ban would be in effect even if the property is sold and thus future owners would have neither insurance nor access to federal reconstruction assistance (Collins 1997). If a federal earthquake insurance is developed, the question is open as to whether a similar 'one time' provision would be made for victims of earthquakes. Removing the so-called federal 'safety net' offered by FEMA grant and loan programs has been suggested as a strategy to increase demand for earthquake insurance and force homeowners to take responsibility for protecting their own property (Comerio *et al.* 1996).

REFERENCES

Abramowitz, M. (1996) *Under Attack, Fighting Back*, New York: Cornerstone.

Aguilar, A. and N. Popovic (1994) 'Lawmaking in the United Nations: The UN Study on Human Rights and the Environment', *Reciel* 3 (4): 12–19.

Aguirre, B. (1994) 'Collective Behavior and Social Movement Theory', in R. Dynes and K. Tierney (eds) *Disasters, Collective Behavior, and Social Organization*, Newark: University of Delaware Press.

Albala-Bertrand, J. (1993) *The Political Economy of Large Natural Disasters*, Oxford: Clarendon Press.

Alesch, D. and W. Petak (1986) *The Politics and Economics of Earthquake Hazard Mitigation: Unreinforced Masonry Buildings in Southern California*, Boulder: Institute of Behavioral Science, University of Colorado.

Alexander, D. (1993) *Natural Disaster*, London: UCL Press.

Ali, M. (1987) 'Women in Famine', in B. Currey and G. Hugo (eds) *Famine as a Geographical Phenomenon*, Dordrecht: D. Reidel.

Anderson, M. and P. Woodrow (1989) *Rising from the Ashes*, Boulder, CO: Westview Press.

Anderson, S. (1996) 'A City Called Heaven: Black Enchantment and Despair in Los Angeles', in A. Scott and E. Soja (eds) *The City: Los Angeles and Urban Theory at the End of the Twentieth Century*, Berkeley: University of California Press.

Aptekar, L. (1994) *Environmental Disasters in Global Perspective*, New York: G.K. Hall.

Archer, M. (1996) *Culture and Agency: The Place of Culture in Social Theory*, revised edition, Cambridge: Cambridge University Press.

Atkinson, P. (1992) *Understanding Ethnographic Texts*, Qualitative Research Methods Series 25. Newbury Park: Sage.

Barton, A. (1970) *Communities in Disaster*, New York: Anchor.

Bates, F. (1997) *Sociopolitical Ecology: Human Systems and Ecological Fields*, New York: Plenum.

Bates, F. and W. Peacock (1993) *Living Conditions, Disasters, and Development: An Approach to Cross-Cultural Comparisons*, Athens: University of Georgia Press.

—— (1987) 'Disasters and Social Change', in R. Dynes, B. DeMarchi, and C. Pelanda (eds) *The Sociology of Disasters*, Milan: Angelli.

Bates, F. and C. Pelanda (1994) 'An Ecological Approach to Disasters', in R. Dynes and K. Tierney (eds) *Disasters, Collective Behavior, and Social Organization*, Newark: University of Delaware Press.

Baudrillard, J. (1995) *The Gulf War Did Not Occur*, translated by P. Patton, Bloomington: Indiana University Press.

—— (1988) *America*, London: Verso.

Beck, U. (1992) *Risk Society: Toward a New Modernity*, London: Sage.

Bee, P. (1984) 'CUP 4145: Will 100 Units of Farmworker Housing be Built in Ventura County' Unpublished MA thesis, California State University, Northridge.

Benthall, J. (1993) *Disaster, Relief, and the Media*, London: I.B. Tauris.

Benton, T. (ed.) (1996) *The Greening of Marxism*, New York: Guilford.

—— (1993) *Natural Relations: Ecology, Animal Rights and Social Justice*, London: Verso.

Berger, P. and T. Luckmann (1967) *The Social Construction of Reality*, Garden City, NY: Anchor.

Bernard, H. (1994) *Research Methods in Anthropology: Qualitative and Quantitative Approaches*, 2nd edn, Thousand Oaks: Sage.

Best, J. (ed.) (1989) *Images of Issues: Typifying Contemporary Social Problems*, New York: Aldine de Gruyter.

Best, S. (1995) *The Politics of Historical Vision*, New York: Guilford.

Best, S. and D. Kellner (1997) *The Postmodern Turn*, New York: Guilford.

—— (1991) *Postmodern Social Theory*, New York: Guilford.

Birkland, T. (1996) 'Natural Disasters as Focusing Events: Policy Communities and Political Response', *Mass Emergencies and Disasters*, 14 (2): 221–244.

Blaikie, P. (1985) *Political Economy of Soil Erosion*, London: Methuen.

Blaikie, P. and H. Brookfield (1987) *Land Degradation and Society*, London: Methuen.

Blaikie, P., T. Cannon, I. Davis, and B. Wisner (1994) *At Risk: Natural Hazards, People's Vulnerability and Disasters*, London: Routledge.

Blocker, T., E. Rochford, and D. Sherkat (1991) 'Political Responses to Natural Hazards: Social Movement Participation following a Flood Disaster', *International Journal of Mass Emergencies and Disasters*, 9 (3): 367–382.

Blumer, H. (1971) 'Social Problems as Collective Behavior', *Social Problems*, 18: 301–313.

Bolin, R. (1994a) 'Postdisaster Shelter and Housing: Social Processes in Response and Recovery', in R. Dynes and K. Tierney (eds) *Disaster, Collective Behavior, and Social Organization*, Newark: University of Delaware Press.

—— (1994b) *Household and Community Recovery after Earthquakes*, Monograph #56. Boulder: Institute of Behavioral Science, University of Colorado.

Bolin, R. and P. Bolton (1986) *Race, Religion, and Ethnicity in Disaster Recovery*, Monograph #42. Boulder: Institute of Behavioral Science, University of Colorado.

Bolin, R. and D. Klenow (1988) 'Older People in Disaster: A Study of Black and White Victims', *International Journal of Aging and Human Development*, 26 (1): 29–43.

Bolin, R. and L. Stanford (1998) 'The Northridge, California Earthquake: Community-Based Approaches to Unmet Recovery Needs', *Disasters*, 22 (1): 21–38,

—— (1991) 'Shelter, Housing, and Recovery: A Comparison of U.S. Disasters.' *Disasters*, 15 (1): 25–34.

Bolin, R., M. Jackson, and A. Crist (in press) 'Gender, Vulnerability, and Disaster: Issues in Theory and Research', in B. Morrow and E. Enarson (eds) *Through Women's Eyes: The Gendered Terrain of Natural Disasters*, New York: Greenwood.

Bolt, B. (1993) *Earthquakes*, New York. W.H. Freeman.

Bolton, P., E. Liebow, and J. Olson (1992) 'Community Context and Uncertainty following a Damaging Earthquake: The Case of Low-Income Latinos in Los Angeles', Paper presented at the Symposium on Policy Issues in the Provision of Post-Earthquake Shelter and Housing, January, Santa Cruz, California.

Booth, D. (ed.) (1995) *Rethinking Social Development*, London: Methuen.

Bottles, S. (1987) *Los Angeles and the Automobile*, Berkeley: University of California Press.

Bourdieu, P. and L. Wacquant (1992) *An Invitation to Reflexive Sociology*, Cambridge: Polity.

Brown, E. (1991) 'Sex and Starvation: Famine in Three Chadian Societies', in R. Downs, D. Kerner, and S. Reyna (eds) *Political Economy of an African Famine*, Philadelphia: Gordon and Breach.

Brown, L. (1996) 'The Acceleration of History', in L. Brown (ed.) *State of the World 1996*, New York: Norton.

Bullard, R. (ed.) (1994) *Unequal Protection: Environmental Justice and Communities of Color*, San Francisco: Sierra Club Books.

—— (ed.) (1993) *Confronting Environmental Racism: Voices from the Crossroads*. Boston: South End Press.

Bullard, R. (1990) *Dumping in Dixie*, Boulder, CO: Westview Press.

Burton, I., R. Kates, and G. White (1993) *The Environment as Hazard*, 2nd edn, New York: Guilford.

Camacho de Schmidt, A. and A. Schmidt (1995) 'Foreword: The Shaking of a Nation', translators' introduction to E. Poniatowska, *Nothing, Nobody: The Voices of the Mexico City Earthquake*, Philadephia, PA: Temple University Press.

Cannon, T. (1994) 'Vulnerability Analysis and the Explanation of "Natural" Disasters', in A. Varley (ed.) *Disasters, Development and Environment*, London: Wiley.

CDI (California Department of Insurance) (1997) 'Homeowners Premium Survey Shows Rates Climbing', Unpublished document, Sacramento: California Department of Insurance.

—— (1996) 'Fair Plan Extension to End May 31', Unpublished document, Sacramento: California Department of Insurance.

CEDC (Cabrillo Economic Development Corporation) (1992) *Special Annual Report: 1981–1992*. Ventura: Cabrillo Economic Development Corporation.

Chambers, R. (1983) *Rural Development: Putting the Last First*, London: Longman.

City of Fillmore (1992) *General Plan: 1988–2010*, Public Document, Fillmore, CA: City of Fillmore.

—— (1994) *City of Fillmore Downtown Specific Plan*, Fillmore, California.

City of Los Angeles (1994) 'Current Status of Damage to the Housing Stock of the City of Los Angeles from the Northridge Earthquake', City of Los Angeles: Los Angeles Housing Department.

City of Santa Clarita (1994a) *1994: A Year in Review*, Unpublished report, Santa Clarita, California.

—— (1994b) *Preliminary After Action Report*, Unpublished report, Santa Clarita, California.

Collins, F. (1997) 'Social and Economic Changes that will Effect Disaster Recovery – What Voluntary Agencies Should Consider', Unpublished report, Richmond: NorCal VOAD.

Comerio, M. (1997) 'The Impact of Housing Losses in the Northridge Earthquake', in *Proceedings of the Northridge Earthquake Conference*, Richmond: California Universities for Research in Earthquake Engineering.

—— (1996) *The Impact of Housing Losses in the Northridge Earthquake: Recovery and Reconstruction Issues*, Berkeley: Center for Environmental Design Research Publications CEDR-14-96 and 15-96.

—— (1995) *The Northridge Earthquake Housing Losses*, Report to the California Governor's Office of Emergency Services. Berkeley, CA: Center for Environmental Design Research, University of California.

Comerio, M., J. Landis, C. Firpo, and J. Monzon (1996) *Residential Earthquake Recovery: Improving California's Post-Disaster Rebuilding Policies and Programs*, Berkeley: California Policy Seminar Brief Series.

Community of Piru (1995) Redevelopment Agency Project Area Sub-Committee Meeting, Unpublished meeting notes, 22 February 1995, Piru, California.

Connell, R. (1993) 'The Big Picture: Masculinities in Recent World History.' *Theory and Society*, 22 (Oct.): 597–623.

Copans, J. (1983) 'The Sahelian Drought: Social Sciences and the Political Economy of Underdevelopment.' in K. Hewitt (ed.) *Interpretations of Calamity from the Perspective of Human Ecology*, London: Allen and Unwin.

Cuny, F. (1983) *Disasters and Development*, New York: Oxford University Press.

Daly, H. and J. Cobb (1989) *For the Common Good: Redirecting the Economy Toward Community, the Environment, and a Sustainable Future*, Boston: Beacon.

Dash, N., W. Peacock, and B. Morrow (1998) 'And the Poor Get Poorer.' in W. Peacock, B. Morrow, and H. Gladwin (eds) *Hurricane Andrew and Miami: Toward a New Socio-Political Ecology of Disasters*, London: Routledge.

Davis, I. and J. Hodson (1982) *Witnesses to Political Violence in Guatemala: The Suppression of a Rural Development Movement*, Boston: Oxfam America.

Davis, M. (1996) 'Water Pirates and the Infinite Suburb', *Capitalism Nature Socialism*, 7 (2): 81–85.

—— (1995) 'How Eden Lost its Garden: A Political History of the L.A. Landscape', *Capitalism Nature Socialism*, 6 (4): 1–30.

—— (1992) *City of Quartz*, New York: Vintage.

—— (1986) *Prisoners of the American Dream*, London: Verso.

Dear, M. (1996) 'In the City Time becomes Visible: Intentionality and Urbanism in Los Angeles, 1781–1991', in A. Scott and E. Soja (eds) *The City: Los Angeles and Urban Theory at the End of the Twentieth Century*, Berkeley: University of California Press.

Demko, G. and M. Jackson (eds) (1995) *Populations at Risk in America: Vulnerable Groups at the End of the Twentieth Century*, Boulder, CO: Westview Press.

Dickens, P. (1996) *Reconstructing Nature: Alienation, Emancipation and the Division of Labor*, London: Routledge.

Dickens, P. and A. Fontana (1994) *Postmodernism and Social Inquiry*, New York: Guilford.

Douglas, M. and A. Wildavsky (1982) *Risk and Culture: An Essay on the Selection of Technical and Environmental Dangers*, Berkeley: University of California Press.

Downs, R., D. Cook, and S. Reyna (eds) (1991) *The Political Economy of an African Famine*, Philadelphia: Gordon and Breach.

Drabek, T. (1994) 'Disaster in Aisle 13 Revisited', in R. Dynes and K. Tierney (eds)

Disasters, Collective Behavior, and Social Organization, Newark: University of Delaware Press.

—— (1989) 'Disasters as Non-Routine Social Problems', *International Journal of Mass Emergencies and Disasters*, 7 (3): 253–264.

—— (1986) *Human System Responses to Disaster*, New York: Springer-Verlag.

Dreze, J. and A. Sen (1990) *Political Economy of Hunger*. Vol. 1. *Entitlement and Well-being*, Oxford: Clarendon Press.

—— (1989) *Hunger and Public Action*, Oxford: Oxford University Press.

Dudasik, S. (1982) 'Unanticipated Repercussions of International Disaster Relief', *Disasters*, 4: 329–338.

Dynes, R. and T. Drabek (1994) 'The Structure of Disaster Research: Its Policy and Disciplinary Implications', *International Journal of Mass Emergencies and Disasters*, 12 (1): 5–24.

Dynes, R. and K. Tierney (eds) (1994) *Disasters, Collective Behavior, and Social Organization*, Newark: University of Delaware Press.

Dynes, R., K. Tierney, and C. Fritz (1994) 'Foreword: The Emergence and Importance of Social Organization: The Contributions of E.L. Quarantelli', in R. Dynes and K. Tierney (eds) *Disasters, Collective Behavior, and Social Organization*, Newark: University of Delaware Press.

Eguchi, R., J. Goltz, and H. Seligson (1997) 'Integrated Real-Time Disaster Information Systems: The Application of New Technologies', Paper presented at the US–Japan Joint Seminar on Civil Infrastructure Systems Research, Honolulu, Hawaii, 28–30 August.

Ehrenreich, B. (1997) 'When Government Gets Mean', *The Nation*, 265 (16): 12–16.

Eldar, R. (1992) 'The Needs of Elderly Persons in Natural Disasters: Observations and Recommendation', *Disasters*, 16 (4): 355–358.

Enarson, E. and B. Morrow (1998) 'A Gendered Perspective: The Voices of Women', in Peacock, W., B. Morrow, and H. Gladwin (eds) *Hurricane Andrew and Miami: Toward a New Socio-Political Ecology of Disasters*, London: Routledge.

Engels, F. (1959) *The Dialectics of Nature*, Moscow: Progress.

Erikson, K. (1994) *A New Species of Trouble*, New York: Norton.

—— (1976) *Everything in its Path*, New York: Simon and Schuster.

Escobar, A. (1996) 'Constructing Nature', in R. Peet and M. Watts (eds) *Liberation Ecologies: Environment, Development, Social Movements*, London: Routledge.

Farganis, S. (1995) *The Social Reconstruction of the Feminine Character*, 2nd edn, New York: Rowan-Littlefield.

Faux, J. (1997) 'The "American Model" Exposed', *The Nation*, 265 (13): 18–22.

FEMA (Federal Emergency Management Agency) (1996) *Status of the Northridge Earthquake Recovery – Sept. 6, 1996*, Pasadena: FEMA/OES Northridge Long-Term Recovery Area Office.

—— (1995) *The Northridge Earthquake One Year Later*, Washington, DC: Federal Emergency Management Agency.

Ferguson, K. (1993) *The Man Question: Visions of Subjectivity in Feminist Theory*, Berkeley: University of California Press.

FitzSimmons, M. and R. Gottlieb (1996) 'Bounding and Binding Metropolitan Space: The Ambiguous Politics of Nature in Los Angeles.' in A. Scott and E. Soja (eds) *The City: Los Angeles and Urban Theory at the End of the Twentieth Century*, Berkeley: University of California Press.

Flavin, C. (1997) 'Climate Change and Storm Damage: The Insurance Costs Keep Rising.' *World Watch*, 10 (1): 10–11.

Form, W. and H. Nosow (1958) *Community in Disaster*, New York: Harper.

Fothergill, A. (1996) 'Gender, Risk, and Disaster.' *International Journal of Mass Emergencies and Disasters*, 14 (1): 33–56.

Foucault, M. (1980) *Power/Knowledge: Selected Interviews and Other Writings*, Brighton: Harvester Press.

Fritz, C. (1961) 'Disasters.' in R. Merton and R. Nisbet (eds) *Contemporary Social Problems*, New York: Harcourt.

Galster, G., C. Herbig, J. Silver, and M. Valera (1995) *Documentation of LA Earthquake Disaster Housing Assistance Policies and Procedures*, Washington, DC: Urban Institute for the US Department of Housing and Urban Development (UI6488).

Gamson, W. and A. Modigliani (1989) 'Media Discourse and Public Opinion: A Constructionist Approach.' *American Journal of Sociology*, 95 (1): 1–37.

GAO (US General Accounting Office) (1997) *Housing and Urban Development: Potential Implications of Legislation Proposing to Dismantle HUD*, Washington, DC: GAO/RCED-97-36.

Garner, A. and W. Huff (1997) 'The Wreck of Amtrak's Sunset Limited: News Coverage of a Mass Transport Disaster.' *Disasters*, 21 (1): 4–19.

Garreau, J. (1991) *Edge City*, New York: Doubleday.

Giddens, A. (1990) *The Consequences of Modernity*, Stanford, CA: Stanford University Press.

—— (1984) *The Constitution of Society*, Berkeley: University of California Press.

Gimenez, M. (1995) 'The Production of Divisions: Gender Struggles under Capitalism.' in A. Callari, S. Cullenberg, and C. Biewener (eds) *Marxism in the Postmodern Age*, New York: Guilford.

Goheen, M. (1991) 'Ideology, Gender, and Change: Social Relations of Production and Reproduction in Nso, Cameroon.' in R. Downs, D. Kerner, and S. Reyna (eds) *The Political Economy of African Famine*, Philadelphia, PA: Gordon and Breach.

Goltz, J. (1994) *The Northridge, California Earthquake of January 17, 1994: General Reconnaissance Report*, Technical Report NCEER-94-0005, Buffalo: National Center for Earthquake Engineering Research.

Gottlieb, R. (1993) *Forcing the Spring*, Washington, DC: Island Press.

—— (1988) *A Life of Its Own: The Politics and Power of Water*, New York: Harcourt.

Gottlieb, R. and M. FitzSimmons (1991) *Thirst for Growth: Water Agencies as Hidden Government in California*, Tucson: University of Arizona Press.

Gould, J., B. Goldman, and K. Millpointer (1990) *Deadly Deceit: Low Level Radiation, High Level Coverup*, New York: Four Wall Eight Windows Press.

Gouldner, A. (1970) *The Coming Crisis of Western Sociology*, New York: Basic Books.

Greenhalgh, S. (1995) 'Anthropology Theorizes Reproduction: Integrating Practice, Political Economic, and Feminist Perspectives.' in S. Greenhalg (ed.) *Situating Fertility*, Cambridge: Cambridge University Press.

Gusfield, J. (1994) 'The Reflexivity of Social Movements: Collective Behavior and Mass Society Theory Revisited.' in E. Larana, H. Johnston, and J. Gusfield (eds) *New Social Movements: From Ideology to Identity*, Philadelphia, PA: Temple University Press.

Haas, J., R. Kates, and M. Bowden (1977) *Reconstruction Following Disaster*, Cambridge: MIT Press.

Hannigan, J. (1995) *Environmental Sociology: A Social Constructionist Perspective*, London: Routledge.

Habermas, J. (1987) *The Theory of Communicative Action, Vol. 2: Lifeworld and System: A Critique of Functionalist Reason*. Boston: Beacon Press.

Hamel, J. (1993) *Case Study Methods*, Qualitative Research Methods Series 32. Newbury Park: Sage.

Haraway, D. (1989) *Primate Visions: Gender, Race and Nature in the World of Modern Science*, London: Routledge.

Harris, S. (1990) *Agents of Chaos: Earthquakes, Volcanoes, and other Natural Disasters*, Missoula: Mountain Press.

Harvey, D. (1996) *Justice, Nature, and the Geography of Difference*, Oxford: Blackwell.

—— (1990) *The Condition of Postmodernity*, Cambridge, MA: Blackwell.

—— (1989) *The Urban Experience*, Oxford: Blackwell.

—— (1985a) *Consciousness and the Urban Experience*, Baltimore: Johns Hopkins University Press.

—— (1985b) *The Urbanization of Capital*, Baltimore: Johns Hopkins University Press.

Hayes, D. (1989) *Behind the Silicon Curtain*, Boston: South End.

Hecht, S. and A. Cockburn (1990) *Fate of the Forest: Developers, Destroyers, and Defenders of the Amazon*, New York: HarperCollins.

Hendrie, B. (1997) 'Knowledge and Power: A Critique of an International Relief Operation.' *Disasters*, 21 (1): 57–76.

Herman, E. and N. Chomsky (1988) *Manufacturing Consent: The Political Economy of the Mass Media*, New York: Pantheon.

Hewitt, K. (1997) *Regions of Risk: A Geographical Introduction to Disasters*, London: Longman.

—— (1995) 'Excluded Perspectives in the Social Construction of Disaster.' *International Journal of Mass Emergencies and Disasters*, 13 (3): 317–340.

—— (ed.) (1983a) *Interpretations of Calamity from the Perspective of Human Ecology*, London: Allen and Unwin.

—— (1983b) 'The Idea of Calamity in a Technocratic Age.' in K. Hewitt (ed.) *Interpretations of Calamity from the Perspective of Human Ecology*, London: Allen and Unwin.

Hiroi, O. (1997) 'Some Problems in the Great Hanshin-Awaji Earthquake in Japan', Paper presented at the NCEER-INCEDE Center to Center Workshop on the Kobe and Northridge Earthquakes, Wilmington, Delaware, 18–20 September.

Hirose, H. (1992) 'Defining Disaster Relief: Disaster Victims and Disaster Relief Administration in the Case of Mt Unzen's Eruption', *International Journal of Mass Emergencies and Disasters*, 10 (2): 281–292.

Horton, J. (1992) 'The Politics of Diversity in Monterey Park, California.' in L. Lamphere (ed.) *Restructuring Diversity*, Chicago: University of Chicago Press.

Horton, S. (1995) 'Rethinking Recycling: The Politics of the Waste Crisis', *Capitalism Nature Socialism*, 6 (1): 1–21.

Hundley, N. (1992) *The Great Thirst: Californians and Water, 1770s–1990s*, Berkeley: University of California Press.

IFRCRCS (International Federation of Red Cross and Red Crescent Societies) (1996) *World Disasters Report 1996*, Oxford: Oxford University Press.

Impact Assessment Incorporated (1990) *The Social, Economic, and Psychological Effects of the Exxon Valdez Oil Spill*, Final report. Anchorage, Alaska.

Jackson, C. (1994) 'Gender Analysis and Environmentalisms.' in M. Redclift and T. Benton (eds) *Social Theory and Global Environments*, Routledge: London.

Jackson, S. (1993) 'Women and the Family.' in D. Richardson and V. Robinson (eds), *Thinking Feminist*, New York: Guilford Press.

Jameson, F. (1991) *Postmodernism, or, The Cultural Logic of Late Capitalism*, Durham, NC: Duke University Press.

Johnston, B. (1994) *Who Pays the Price? The Sociocultural Context of the Environmental Crisis*, Washington, DC: Island Press.

Kahrl, W. (1982) *Water and Power*, Berkeley: University of California Press.

Keen, D. (1994) *The Benefits of Famine: The Political Economy of Famine and Relief in Southwestern Sudan, 1983–1989*, Princeton, NJ: Princeton University Press.

Kent, R. (1987) *Anatomy of Disaster Relief*, London: Pinter.

Kirby, A. (1990) 'Things Fall Apart: Risks and Hazards in their Social Setting.' in A. Kirby (ed.) *Nothing to Fear: Risks and Hazards in American Society*, Tucson: University of Arizona Press.

Knorr Cetina, K. (1992) 'The Couch, the Cathedral, and the Laboratory: On the Relation between the Experiment and the Laboratory in Science.' in A. Pickering (ed.) *Science as Practice and Culture*, Chicago: University of Chicago Press.

Kreimer, A. and E. Echeverria (1991) 'Case Study: Housing Reconstruction in Mexico City.' in A. Kreimer and M. Munasinghe (eds) *Managing Natural Disasters and the Environment*, Washington, DC: World Bank.

Kreimer, A. and M. Munasinghe (eds) (1992) *Environmental Management and Urban Vulnerability*, World Bank Discussion Paper No. 168. Washington, DC: World Bank.

Kreps, G. (1995) 'Disaster as Systemic Event and Social Catalyst: A Clarification of the Subject Matter.' *International Journal of Mass Emergencies and Disasters*, 13 (3): 255–284.

Kreps, G. and S. Bosworth (1994) *Organizing, Role Enactment, and Disaster: A Structural Theory*, Newark: University of Delaware Press.

Kreps, G. and T. Drabek (1996) 'Disasters as Non-Routine Social Problems.' *International Journal of Mass Emergencies and Disasters*, 14 (2): 129–154.

Kroll-Smith, S. and S. Couch (1991a) *The Real Disaster is Above Ground: A Mine Fire and Social Conflict*, Lexington: University of Kentucky Press.

—— (1991b) 'What is a Disaster? An Ecological-Symbolic Approach to Resolving the Definitional Debate.' *International Journal of Mass Emergencies and Disasters*, 9 (3): 355–366.

Kuhn, T. (1970) *The Structure of Scientific Revolutions*, 2nd edn, Chicago: University of Chicago Press.

LAHD (Los Angeles Housing Department) (1995) 'Rebuilding Communities: Recovering from the January 17, 1994 Northridge Earthquake', Unpublished report, Los Angeles: City of Los Angeles.

Laird, R. (1991) 'Ethnography of a Disaster', Unpublished MA thesis, San Francisco: San Francisco State University.

Lambright, W. (1985) 'Earthquake Prediction and the Governmental Process.' *Policy Studies Review*, 4 (4): 716–724.

Lamphere, L. (1992) 'Introduction: The Shaping of Diversity.' in L. Lamphere (ed.) *Structuring Diversity*, Chicago: University of Chicago Press.

Lemert, C. (1995) *Sociology after the Crisis*, Boulder, CO: Westview Press.

Lindell, M. (1994) 'Perceived Characteristics of Environmental Hazards.' *International Journal of Mass Emergencies and Disasters*, 12 (3): 303–326.

Lo, C. (1990) *Small Property versus Big Government*, Berkeley: University of California Press.

Los Angeles Times (1994) *4: 31: Images of the 1994 Los Angeles Earthquake*, Los Angeles: Los Angeles Times Syndicate.

Loughlin, K. (1995) 'Rehabilitating Science, Imagining Bhopal.' pp. 277–302 in G. Marcus (ed.) *Late Editions 2: Technoscientific Imaginaries*, Chicago: University of Chicago Press.

Lutz, C. (1995) 'The Gender of Theory.' in R. Behar and D. Gordon (eds) *Women Writing Culture*, Berkeley: University of California Press.

Mann, A. (1979) *The Field Act and California Schools*, Sacramento, CA: Seismic Safety Commission.

Maskrey, A. (1994) 'Disaster Mitigation as a Crisis of Paradigms: Reconstruction after the Alto Mayo Earthquake, Peru', in A. Varley (eds.) *Disasters, Development and Environment*, Chichester, UK: Wiley.

—— (1993) *Los Desastros NO Son Naturales*, Bogota: La Red/ITDG.

Mate, K. (1982) '"For Whites Only": The South Bay Perfects Racism for the '80s.' *LA Weekly*, (August): 11–14.

May, P. (1985) *Recovering from Catastrophes: Federal Disaster Relief Policy and Politics*, Westport, CT: Greenwood Press.

May, P. and T. Birkland, (1994) 'Earthquake Risk Reduction: An Examination of Local Regulatory Effects', *Environmental Management*, 18: 923–939.

McAdam, D. (1997) 'The Classical Model of Social Movements Examined.' in S. Buechler and K. Cylke (eds) *Social Movements: Perspectives and Issues*, Mountain View, CA: Mayfield.

McBane, M. (1995) 'The Role of Gender in Citrus Employment: A Case Study of Recruitment, Labor, and Housing Patterns at the Limoneira Company, 1893–1940', *California History*, 74 (1): 68–81.

McWilliams, C. (1976) *California: The Great Exception*, Santa Barbara, CA: Peregrine Smith.

Menchaca, M. (1995) *The Mexican Outsiders: A Community History of Marginalization and Discrimination in California*, Austin: University of Texas Press.

Merchant, C. (1992) *Radical Ecology*, New York: Routledge.

Meyers, B. (1991) 'Disaster Study of War.' *Disasters*, 15 (4): 318–330.

Mileti, D. (1997) 'Designing Disasters: Determining our Future Vulnerability', *Natural Hazards Observer*, 22 (2): 1–3.

Mileti, D. and C. Fitzpatrick (1993) *The Great Earthquake Experiment*, Boulder, CO: Westview Press.

Mileti, D. and J. Sorensen (1990) *Communication of Emergency Public Warnings: A Social Science Perspective and State-of-the-Art Assessment*, Washington, DC: Federal Emergency Management Agency.

Mileti, D., J. Hutton, and J. Sorenson (1981) *Earthquake Prediction Response and Options for Public Policy*, Boulder: Institute of Behavioral Science, University of Colorado.

Miller, G. (1981) *Cities by Contract: The Politics of Municipal Incorporation*, Cambridge, MA: MIT Press.

Milton, K. (1996) *Enivronmentalism and Cultural Theory*, London: Routledge.

Mishler, E. (1991) 'Representing Discourse: The Rhetoric of Transcription', *Journal of Narrative and Life History* 1(4): 255–280.

Mitchell, J. (1990) 'Human Dimensions of Environmental Hazards.' in A. Kirby (ed.) *Nothing to Fear: Risks and Hazards in American Society*, Tucson: University of Arizona Press.

Mittler, E. and R. Stallings (1997) 'Neighborhood Retention after the Northridge Earthquake or Expectations Unfulfilled: What Was and What Could Have Been', in *Proceedings of the Northridge Earthquake Conference*, Richmond: California Universities for Research in Earthquake Engineering.

Moore, D. (1996) 'Marxism, Culture, and Political Ecology: Environmental Struggles in Zimbabwe's Eastern Highlands.' in R. Peet and M. Watts (eds) *Liberation Ecologies: Environment, Development, Social Movements*, London: Routledge.

Morgan, David L. (1988) *Focus Groups as Qualitative Research*, Qualitative Research Methods Series 16. Newbury Park: Sage.

Morrow, B. and E. Enarson (1996) 'Hurricane Andrews through Women's Eyes: Issues and Recommendations.' *International Journal of Mass Emergencies and Disasters*, 14 (1) : 5–22.

Morrow, B. and W. Peacock (1998) 'Disasters and Social Change: Hurricane Andrew and the Reshaping of Miami', in W. Peacock, B. Morrow, and H. Gladwin (eds) *Hurricane Andrew and Miami: Toward a New Socio-Political Ecology of Disasters*, London: Routledge.

Neal, D. (1997) 'Reconsidering the Phases of Disaster', *International Journal of Mass Emergencies and Disasters*, Vol. 15 (2): 239–264.

Neal, D. and B. Phillips (1995) 'Effective Emergency Management: Reconsidering the Bureaucratic Approach.' *Disasters*, 19 (4): 327–337.

Nelson, H. (1983) *The Los Angeles Metropolis*, Dubuque, IA: Candle/Hunt.

NAS (National Academy of Sciences), Committee on the Alaska Earthquake (eds) (1970) *The Great Alaska Earthquake of 1964: Human Ecology*, Washington, DC: NAS.

Nigg, J. (1997) 'Emergency Response Following the 1994 Northridge Earthquake: Intergovernmental Coordination Issues', in *Proceedings of the Northridge Earthquake Research Conference*, Richmond: California Universities for Research in Earthquake Engineering.

NRC (National Research Council) (1994) *Practical Lessons from the Loma Prieta Earthquake*, Washington, DC: National Academy Press.

O'Brien, J. and E. Gruenbaum (1991) 'A Social History of Food, Famine, and Gender in Twentieth-Century Sudan.' in R. Downs, D. Kerner, and S. Reyna (eds) *The Political Economy of African Famine*, Philadelphia: Gordon and Breach.

O'Connor, J. (1994) 'Is Sustainable Capitalism Possible?', in M. O'Connor (ed.) *Is Capitalism Sustainable? Political Economy and the Politics of Ecology*, New York: Guilford.

O'Connor, M. (ed.) (1994) *Is Capitalism Sustainable? Political Economy and the Politics of Ecology*, New York: Guilford.

OES (California Governor's Office of Emergency Services) (1997) *The Northridge Earthquake of January 17, 1994: Report of Data Collection and Analysis, Part B:*

Analysis and Trends, Irvine and Pasadena: EQE International and Office of Emergency Services.

—— (1995) *The Northridge Earthquake of January 17, 1994: Preliminary Report of Data Collection and Analysis, Part A: Damage and Inventory Data*, Irvine and Pasadena: EQE International and Office of Emergency Services.

Office of Earthquakes and Natural Hazards (1993) *Improving Earthquake Mitigation: Report to Congress*, Washington, DC: Federal Emergency Management Agency.

Office of Foreign Disaster Assistance (1990) *Disaster History*, Washington, DC: Office of Foreign Disaster Assistance, Agency for International Development.

Oliver-Smith, A. (1996) 'Anthropological Research on Hazards and Disasters.' *Annual Review of Anthropology*, 25: 303–328.

—— (1994) 'Peru's Five Hundred Year Earthquake. Vulnerability in Historical Context.' in A. Varley (ed.) *Disasters, Development and Environment*, London: Wiley.

—— (1986) *The Martyred City: Death and Rebirth in the Peruvian Andes*, Albuquerque: University of New Mexico Press.

—— (1982) 'Here there is Life: The Social and Cultural Dynamics of Successful Resistance to Resettlement in Post-Disaster Peru.' in A. Hansen and A. Oliver-Smith (eds) *Involuntary Migration and Resettlement: The Problems and Responses of Dislocated People*, Boulder, CO: Westview Press.

Olson, R. (1985) 'Political Economy of Life Safety: The City of Los Angeles and the "Hazardous Structure Abatement".' *Policy Studies Review*, 4 (4): 670–679.

Olson, R. and A. Drury (1997) 'Untherapeutic Communities: A Cross-National Analysis of Post-Disaster Political Unrest', *International Journal of Mass Emergencies and Disasters*, 15 (2): 221–238.

Olson, R., C. Holland, H. Miller, W. Lambright, H. Lagorio, and C. Treseder (1988) *To Save Lives and Protect Property: A Policy Assessment of Federal Earthquake Activities, 1964–1987*, Sacramento, CA: VSP Associates.

Ong, P. and E. Blumenberg (1996) 'Income and Racial Inequality in Los Angeles.' in A. Scott and E. Soja (eds) *The City: Los Angeles and Urban Theory at the End of the Twentieth Century*, Berkeley: University of California Press.

Outland, C. (1977) *Man-Made Disaster: The Story of the St. Francis Dam*, Glendale, CA: Arthur Clark Publishers.

Palerm, J. (1991) *Farm Labor Needs and Farm Workers in California, 1970–1989*, Unpublished report, California Agricultural Studies, Employment Development Department.

Palerm, J. and J. Urquiola (1993) 'A Binational System of Agricultural Production: The Case of the Mexican Bajío and California', in *Mexico and the United States: Neighbors in Crisis*, Conference Proceedings, University of California Institute for Mexico and the United States (UC MEXUS), California.

Palinkas, L., M. Downs, J. Petterson, and J. Russell (1993) 'Social, Cultural, and Psychological Impacts of the Exxon Valdez Oil Spill.' *Human Organization*, 52: 1–13.

Palm, R. (1990) *Natural Hazards: An Integrative Theory for Research and Planning*, Baltimore, MD: Johns Hopkins University Press.

Palm, R. (1995) *Earthquake Insurance: A Longitudinal Study of California Homeowners*, Boulder, CO: Westview Press.

Palm, R. and M. Hodgson (1992) *After a California Earthquake*, Geography Research Paper #233. Chicago: University of Chicago.

Palm, R., M. Hodgson, D. Blanchard, and D. Lyons (1990) *Earthquake Insurance in California: Environmental Policy and Individual Decision Making*, Boulder, CO: Westview Press.

Peacock, W., B. Morrow, and H. Gladwin (eds) (1998) *Hurricane Andrew and Miami: Toward A New Socio-Political Ecology of Disasters*, London: Routledge.

Peacock, W. and K. Ragsdale (1998) 'Social Systems, Ecological Networks and Disasters: Toward a Socio-Political Ecology of Disasters.' in W. Peacock, B. Morrow, and H. Gladwin (eds) *Hurricane Andrew and Miami: Toward a New Socio-Political Ecology of Disasters*, London: Routledge.

Peet, R. and M. Watts (1996) 'Liberation Ecology: Development, Sustainability, Environment in an Age of Market Triumphalism.' in R. Peet and M. Watts (eds) *Liberation Ecologies: Environment, Development, Social Movements*, London: Routledge.

Peirce, N. and R. Guskind (1993) *Breakthroughs: Re-creating the American City*, New Brunswick, NJ: Center for Urban Policy Research, Rutgers University.

Pelanda, C. (1982) 'Disaster and Order: Theoretical Problems in Disaster Research.' Paper presented at the Tenth World Congress of Sociology, Mexico City.

Perry, R. (1987) 'Disaster Preparedness and Response Among Minority Citizens.' in R. Dynes, B. DeMarchi, and C. Pelanda (eds) *Sociology of Disasters*, Milan: Angelli.

—— (1985) *Comprehensive Emergency Management*, Greenwich: JAI Press.

Petak, W. and A. Atkinson (1985) 'Natural Hazard Losses in the United States: A Public Problem', *Policy Studies Review*, 4 (4): 662–669.

Pettiford, L. (1995) 'Toward a Redefinition of Security in Central America: The Case of Natural Disasters', *Disasters*, 19 (2): 148–155.

Phillips, B. (1993) 'Cultural Diversity in Disaster: Shelter, Housing, and Long-Term Recovery', *International Journal of Mass Emergencies and Disasters*, 11(1): 99–110.

—— (1991) *Post-Disaster Sheltering and Housing of Hispanics, the Elderly and the Homeless*, Final report to the National Science Foundation, Dallas, Texas: Southern Methodist University.

Picou, J., D. Gill, and M. Cohen (eds) (1997) *The Exxon Valdez Disaster*, Dubuque, IA: Kendall/Hunt.

Picou, J., D. Gill, C. Dyer, and E. Curry (1992) 'Disruption and Stress in an Alaskan Fishing Community: Initial and Continuing Impacts of the Exxon-Valdez Oil Spill.' *Industrial Crisis Quarterly*, 6: 235–257.

Pirages, D. (1994) 'Sustainability as an Evolving Process.' *Futures*, 26 (2): 197–205.

Ploughman, P. (1997) 'Disasters, the Media and Social Structures: A Typology of Credibility Hierarchy Persistence on Newspaper Coverage of the Love Canal and Six Other Disasters.' *Disasters*, 21 (2): 118–137.

—— (1995) ' The American Print News Media "Construction" of Five Natural Disasters.' *Disasters*, 19 (4): 308–326.

Plumwood, V. (1993) *Feminism and the Mastery of Nature*, London: Routledge.

Poniatowska, E. (1995) *Nothing, Nobody: The Voices of the Mexico City Earthquake*, translated by A. Camacho de Schmidt and A. Schmidt, Philadelphia, PA: Temple.

Pred, A. and M. Watts (1992) *Reworking Modernity: Capitalism and Symbolic Discontent*, New Brunswick, NJ: Rutgers University Press.

Quarantelli, E. (1997) 'Ten Criteria for Evaluating Emergency Management of Community Disasters.' *Disasters*, 21 (1): 39–56.

—— (1995) 'What is a Disaster?' *International Journal of Mass Emergencies and Disasters*, 13 (3): 221–230.

—— (1994) 'Disaster Studies: The Consequences of the Historical Use of a Sociological Approach in the Development of Research', *International Journal of Mass Emergencies and Disasters*, 12 (1): 25–50.

—— (1992) 'Can and Should Social Science Disaster Research Knowledge from Developed Societies be Applied to Developing Societies?' *Asia-Pacific Journal of Rural Development*, 2: 1–14.

—— (1991) 'Disaster Research Center, University of Delaware.' *Disasters*, 15 (3): 274–277.

—— (1990) 'Disaster Prevention and Mitigation in Lada.' Paper presented at Colloquium on the Environment and Natural Disaster Management, Washington, DC: World Bank.

—— (1989) 'Conceptualizing Disaster from a Sociological Perspective.' *International Journal of Mass Emergencies and Disasters*, 7: 243–251.

—— (1987) 'Disaster Studies: An Analysis of the Social Historical Factors Affecting the Development of Research in the Area.' *International Journal of Mass Emergencies and Disasters*, 5: 285–310.

Quarantelli, E. and R. Dynes (1989) 'Reconstruction in the Context of Recovery: Some Thoughts on the Alaska Earthquake.' Paper presented at conference on Reconstruction After Urban Earthquakes, Buffalo: National Center for Earthquake Engineering Research. State University of New York.

Reisner, M. (1993) *Cadillac Desert*, New York: Penguin.

Renteria, H. (1990) 'Loma Prieta Earthquake: Social Response to Housing and Food Requirements.' Paper presented at the conference on Putting the Pieces Together: The Loma Prieta Earthquake One Year Later, October, San Francisco, California.

Riessman, C. (1993) *Narrative Analysis*, Qualitative Research Methods Series 30. Newbury Park: Sage.

Robinson, S., Y. Hernandez, R. Castrejon, and H. Bernard (1986) 'The Mexico City Earthquake.' in A. Oliver-Smith (ed.) *Natural Disasters and Cultural Responses*, Williamsburg: College of William and Mary.

Rocco, R. (1996) 'Latino Los Angeles: Reframing Boundaries/Borders.' in A. Scott and E. Soja (eds) *The City: Los Angeles and Urban Theory at the End of the Twentieth Century*, Berkeley: University of California Press.

Rocheleau, D., B. Thomas-Slayter, and E. Wangari (1996) 'Gender and Environment', in D. Rocheleau, B. Thomas-Slayter, and E. Wangari (eds) *Feminist Political Ecology*, London: Routledge.

Rodman, M. (1992) 'Empowering Place: Multilocality and Multivocality.' *American Anthropologist*, 94 (3): 640–656.

Rogers, J. (1995) 'A Man, a Dam, and a Disaster: Mulholland and the St. Francis Dam', in D. Nunis (ed.) *The St. Francis Dam Disaster Revisited*, Los Angeles: Historical Society of Southern California/Ventura County Museum of History and Art.

Rosaldo, R. (1993) *Culture and Truth*, Boston, MA: Beacon.

Rossi, I. (1993) *Community Reconstruction Following an Earthquake*, New York: Praeger.

Rossi, P., J. Wright, and E. Weber-Burton (1982) *Natural Hazards and Public Choice: The State and Local Politics of Hazard Mitigation*, New York: Academic Press.

Roth, R. and T. Van (1997) *California Earthquake Zoning and Probable Maximum Loss Evaluation Program*, Los Angeles: California Department of Insurance.

Rubin, C. and R. Palm (1987) 'National Origin and Earthquake Response: Lessons from the Whittier Narrows Earthquake of 1987', *International Journal of Mass Emergencies and Disasters*, 5 (3): 347–355.

Sachs, A. (1996) 'Upholding Human Rights and Environmental Justice', in *State of the World, 1996*, New York: Norton.

SECURE (Santa Clarita Educated Communities United in Response to Emergencies) (1992a) *Emergency Preparedness Planning Guide for Business and Industry*, Unpublished report, Santa Clarita, California.

—— (1992b) *A Guide to Being Secure*, Unpublished guide, Santa Clarita, California.

Sassen, S. (1992) 'Why Migrations?' *Report on the Americas*, 26 (1): 14–15.

—— (1988) *The Mobility of Labor and Capital: A Study in International Investment and Labor Flow*, New York: Cambridge University Press.

Scanlon, J. (1994a) 'EMS in Halifax after the 6 December 1917 Explosion: Testing Quarantelli's Theories with Historical Data', in R. Dynes and K. Tierney (eds) *Disasters, Collective Behavior, and Social Organization*, Newark: University of Delaware Press.

—— (1994b) 'The Role of EOCs in Emergency Management: A Comparison of Canadian and American Experience.' *International Journal of Mass Emergencies and Disasters*, 12 (1): 51–76.

Schnaiberg, A. and K. Gould (1994) *Environment and Society: The Enduring Conflict*, New York: St Martins.

Schroeder, R. and M. Watts (1991) 'Struggling over Strategies, Fighting over Food.' *Rural Sociology and Development*, 5: 45–72.

Schulte, P. (1991) 'The Politics of Disaster: An Examination of Class and Ethnicity in the Struggle for Power following the 1989 Loma Prieta Earthquake in Watsonville, California', Unpublished MA thesis, Sacramento: California State University.

Scott, A. (1996) 'High-Technology Industrial Development in the San Fernando Valley and Ventura County: Observations on Economic Growth and the Evolution of Urban Form.' in A. Scott and E. Soja (eds) *The City: Los Angeles and Urban Theory at the End of the Twentieth Century*, Berkeley: University of California Press.

Scott, A. and E. Soja (eds) (1996) *The City: Los Angeles and Urban Theory at the End of the Twentieth Century*, Berkeley: University of California Press.

Scott, J. (1990) *Domination and the Arts of Resistance*, New Haven, CT: Yale University Press.

Sen, A. (1988) 'Family and Food. Sex Bias in Poverty.' in T. Srinivasan and P. Bardhan (eds) *Rural Poverty in South Asia*, New York: Columbia University Press.

—— (1981) *Famine and Poverty*, London: Oxford University Press.

Shah, H. (1995) 'Scientific Profiles of "The Big One".' *Natural Hazards Observer*, 20 (2): 1–2.

Shipton, P. (1990) 'African Famines and Food Security: Anthropological Perspectives.' *Annual Review of Anthropology*, 19: 353–394.

Shiva, V. (1989) *Staying Alive*, London: Zed.

Shkilnyk, A. (1985) *A Poison Stronger than Love: The Destruction of an Ojibwa Community*, New Haven, CU: Yale University Press.

Shoaf, K. (1997) 'Patterns of Injury Following the Northridge Earthquake', Paper presented at the NCEER-INCEDE Center to Center Workshop on the Kobe and Northridge Earthquakes, Wilmington, Delaware, 18–20 September.

Sibley, D. (1995) *Geographies of Exclusion: Society and Difference in the West*, London: Routledge.

Slovic, P. (1986) 'Informing and Educating the Public about Risk.' *Risk Analysis*, 6: 280–285.

Smith, J. and I. Wallerstein (eds) (1992) *Creating and Transforming Households*, Cambridge: Cambridge University Press.

Smith, K. (1996) *Environmental Hazards: Assessing Risks and Reducing Disaster*, London: Routledge.

Soja, E. (1996a) *Thirdspace*, London: Blackwell.

—— (1996b) 'Los Angeles 1965–1992: From Crisis-Generated Restructuring to Restructuring-Generated Crisis.' in A. Scott and E. Soja (eds) *The City: Los Angeles and Urban Theory at the End of the Twentieth Century*, Berkeley: University of California Press.

—— (1989) *Postmodern Geographies*, London: Verso.

Soja, E. and A. Scott (1996) 'Introduction to Los Angeles: City and Region.' in A. Scott and E. Soja (eds) *The City: Los Angeles and Urban Theory at the End of the Twentieth Century*, Berkeley: University of California Press.

Soper, K. (1996) 'Greening Prometheus: Marxism and Ecology.' in T. Benton (ed.) *The Greening of Marxism*, New York: Guilford.

Southwick, C. (1996) *Global Ecology in Human Perspective*, New York: Oxford University Press.

Spector, M. and J. Kitsuse (1987) *Constructing Social Problems*, Hawthorne, NY: Aldine de Gruyter.

SSC (Seismic Safety Commission) (1995) *Northridge Earthquake: Turning Loss to Gain*, Sacramento, CA: SSC Report No. 95-01.

Stallings, R. (1996) 'The Northridge Earthquake Ghost Towns', Final report to the National Science Foundation, Los Angeles: School of Public Administration, University of California, Los Angeles.

—— (1995) *Promoting Risk: Constructing the Earthquake Threat*, New York: Aldine de Gruyter.

—— (1994) 'Collective Behavior Theory and the Study of Mass Hysteria.' in R. Dynes and K. Tierney (eds) *Disasters, Collective Behavior, and Social Organization*, Newark: University of Delaware Press.

—— (1991) 'Disasters as Social Problems? A Dissenting View.' *International Journal of Mass Emergencies and Disasters*, 9 (1): 69–74.

Stoffle, R., M. Traugott, J. Stone, P. McIntyre, F. Jensen, and C. Davidson (1991) 'Risk Perception Mapping: Using Ethnography to Define the Locally Affected Population for a Low-Level Radioactive Waste Storage Facility in Michigan.' *American Anthropologist*, 93 (3): 611–635.

Stolarski, N. (1991) 'The Housing Reconstruction Program: Housing Reconstruction after the Earthquakes of 1985 in Mexico City', Paper presented at the Earthquake Recovery and Reconstruction Conference, May 30–31, San Francisco, California.

Subervi-Vélez, F., M. Denney, A. Ozuna, C. Quintero, and J.-V. Palerm (1992) 'Communicating with California's Spanish Speaking Populations: Assessing the Role of the Spanish-Language Broadcast Media and Selected Agencies in

Providing Emergency Services.' *CPS Report*, Berkeley: California Policy Seminar, University of California.

Susman, P., P. O'Keefe, and B. Wisner (1983) 'Disasters, a Radical Interpretation.' in K. Hewitt (ed.) *Interpretations of Calamity from the Viewpoint of Human Ecology*, London: Allen and Unwin.

Szasz, A. (1994) *Ecopopulism: Toxic Waste and the Movement for Environmental Justice*, Minneapolis: University of Minnesota Press.

Tierney, K. (1996) 'Social Aspects of the Northridge Earthquake.' pp. 255–262 in M. Woods and R. Seiple (eds) *The Northridge, California Earthquake of 17 January, 1994*, Sacramento: California Department of Conservation, Division of Mines and Geology.

—— (1989) 'Improving Theory and Research in Hazard Mitigation: Political Economy and Organizational Perspectives.' *International Journal of Mass Emergencies and Disasters*, 7 (3) : 367–396.

Tierney, K. and J. Dahlhamer (1997) 'Business Disruption, Preparedness and Recovery: Lessons from the Northridge Earthquake.' in *Proceedings of the Northridge Earthquake Research Conference*, Richmond: California Universities for Research in Earthquake Engineering.

Tierney, K. and J. Goltz (1996) 'Emergency Response: Lessons Learned from the Kobe Earthquake', Unpublished report, Newark: Disaster Research Center, University of Delaware.

Tobin, G. and B. Montz (1997) *Natural Hazards: Explanation and Integration*, New York: Guilford.

Tobriner, S. (1988) 'The Mexico Earthquake of September 19, 1985: Past Decisions, Present Danger: an Historical Perspective on Ecology and Earthquakes in Mexico City', *Earthquake Spectra*, 4 (3): 469–479.

Torry, W. (1986) 'Economic Development, Drought, and Famine: Some Limitations of Dependency Explanations.' *GeoJournal*, 12: 5–18.

—— (1979) 'Anthropological Studies in Hazardous Environments: Past Trends and New Horizons.' *Current Anthropology*, 20: 517–541.

Triem, J. (1985) *Ventura County: Land of Good Fortune*, Chatsworth, CA: Windsor Publications.

Turner, R. and L. Killian (1987) *Collective Behavior*, Englewood Cliffs, NJ: Prentice-Hall.

Tyler, M. (1997a) 'Demolitions in Fillmore after the Northridge Earthquake', in *Proceedings of the Northridge Earthquake Research Conference*, Richmond: California Universities for Research in Earthquake Engineering.

—— (1997b) 'An Evaluation of the Los Angeles Recovery and Reconstruction Plan after the Northridge Earthquake', in *Proceedings of the Northridge Earthquake Research Conference*, Richmond: California Universities for Research in Earthquake Engineering.

UNDP (United Nations Development Programme) (1990) *Human Development Report 1990*, New York: United Nations.

US Bureau of the Census (1992) *Poverty in the United States: 1964–1989*, Current Population Reports, Series P–60, no. 175. Washington, DC: US Government Printing Office.

USGS (US Geological Survey) (1996) *USGS Response to an Urban Earthquake: Northridge '94*, Open File Report 96-263, Denver: USGS.

US Senate (1995) *Federal Disaster Assistance*, Bipartisan Task Force on Funding Disaster Relief, Washington, DC: US Government Printing Office, Document No. 104-4.

Varley, A. (1994) 'The Exceptional and the Everyday: Vulnerability Analysis in the International Decade for Natural Disaster Reduction', in A. Varley (ed.) *Disasters, Development and Environment*, London: Wiley.

Vaughan, M. (1987) *The Story of an African Famine*, London: Cambridge University Press.

Vaupel, S. (1992) *A Study of Women Agricultural Workers in Ventura County, California*, Report to the County, Ventura: Committee on Women in Agriculture, Ventura County.

Ventura County (1996) *Piru Community Enhancement Plan*, Unpublished report, Ventura County Supervisors, Ventura, California.

—— (1994) 'Public Social Services Agency Disaster Recovery Plan', Unpublished report, Ventura, California.

—— (1987) *Cultural Heritage Survey*, Unpublished report, Ventura County, General Services Agency, Recreation Services, Ventura, California.

Ventura County Department of Mental Health (1995) 'Project COPE, Crisis Counseling Program, Final Report, May 18, 1994 to July 28, 1994', Unpublished report, Ventura County Department of Mental Health, Ventura, California.

Wallerstein, I. (1992) 'America and the World, Today, Yesterday, and Tomorrow', *Theory and Society*, 21: 1–28.

Wallrich, B. (1996) 'The Evolving Role of Community-Based Organizations in Disaster Recovery.' *Natural Hazards Observer*, 21 (2) : 12–13.

Wasserman, H. and N. Solomon (1982) *Killing our Own*, New York: Delta.

Waterstone, M. (1992) 'Introduction: The Social Genesis of Risks and Hazards.' in M. Waterstone (ed.) *Risk and Society: The Interaction of Science, Technology, and Public Policy*, Boston: Kluwer Academic.

Watts, M. (1992) 'Capitalisms, Crises, and Cultures I: Notes toward a Totality of Fragments.' in A. Pred and M. Watts (eds) *Reworking Modernity: Capitalism and Symbolic Discontent*, New Brunswick: Rutgers University Press.

—— (1991) 'Heart of Darkness: Reflections on Famine and Starvation in Africa.' in R. Downs, D. Kerner, and S. Reyna (eds) *Political Economy of an African Famine*, Philadelphia: Gordon and Breach.

—— (1983) 'On the Poverty of Theory: Natural Hazards Research in Context', in K. Hewitt (ed.) *Interpretations of Calamity from the Perspective of Human Ecology*, London: Allen and Unwin.

(WCED) (World Commission on Environment and Development) (1987) *Our Common Future*, New York: Oxford University Press.

White, G. (1945) *Human Adjustment to Floods: A Geographic Approach to the Flood Problem in the United States*, Research Paper 29. Department of Geography, University of Chicago.

White, G. and J. Haas (1975) *Assessment of Research on Natural Hazards*, Cambridge, MA: MIT Press.

Whyte, A. (1986) 'From Hazard Perception to Human Ecology.' in R. Kates and I. Burton (eds) *Geography, Resources, and Environment*: Vol. 2 *Themes from the Work of Gilbert F. White*, Chicago: University of Chicago Press.

Whyte, W. (1958) 'Urban Sprawl.' *Fortune*, 57 (January): 302.

Wiest, R., J. Mocellin, and D. Motisi (1992) *The Needs of Women and Children in Disaster and Emergencies*, Winnipeg, Disaster Research Unit, University of Manitoba. Winnipeg.

Wiggins, E. (1996) 'Letter to the Editor.' *Natural Hazards Observer*, 20 (4): 5.

Wijkman, A. and L. Timberlake (1988) *Natural Disasters: Acts of God or Acts of Man?* Santa Cruz: New Society Publishers.

Winchester, P. (1992) *Power, Choice and Vulnerability: A Case Study in Disaster Mismanagement in South India, 1977–1988*, London: James and James.

Wisner, B. (1993) 'Disaster Vulnerability: Scale, Power and Daily Life', *GeoJournal*, 30 (2): 127–144.

Wolch, J. (1996) 'From Global to Local: The Rise of Homelessness in Los Angles during the 1980s.' in A. Scott and E. Soja (eds) *The City: Los Angeles and Urban Theory at the End of the Twentieth Century*, Berkeley: University of California Press.

—— (1995) 'Inside/Outside: The Dialectics of Homelessness', in G. Demko and M. Jackson (eds) *Populations at Risk in America: Vulnerable Groups at the End of the Twentieth Century*, Boulder, CO: Westview Press.

Wolf, D. (1991) 'Does Father Know Best? A Feminist Critique of Household Strategy Research.' *Research in Rural Sociology and Development*, 5: 29–44.

Wong, D. (1984) 'The Limits of Using the Household as a Unit of Analysis.' in J. Smith, I. Wallerstein, and H. Evers (eds) *Households and the World Economy*. Beverly Hills, CA: Sage.

Woods, E. (1995) *Democracy against Capitalism*, Cambridge: Cambridge University Press.

Wright, W. (1995) *Wild Knowledge: Science, Lanugage, and Social Life in a Fragile Environment*, Minneapolis: University of Minnesota Press.

Yamazaki, F., N. Yamaguchi, and K. Wakamatsu (1997) 'GIS Analysis of Building Damage Due to the 1995 Kobe Earthquake', Paper presented at the NCEER-INCEDE Center to Center Workshop on the Kobe and Northridge Earthquakes, Wilmington, DE, 18–20 September.

Yanarella, E. and R. Levine (1992) 'Does Sustainable Development Lead to Sustainability?' *Futures*, 24 (10): 759–774.

Zaman, M. (1989) 'The Social and Political Context of Adjustment to Riverbank Erosion Hazard and Population Resettlement in Bangladesh.' *Human Organization*, 48: 196–205.

Zinn, H. (1990) *A People's History of the United States*, New York: Harper.

NAME INDEX

Abramowitz, M. 63, 228
Aguilar, A. 231
Aguirre, B. 62
Albala-Bertrand, J. 37, 130, 213–14, 221, 224
Alesch, D. 80, 81
Alexander, D. 24
Ali, M. 41
Anderson, M. 8, 19, 43, 57, 59, 220, 224
Anderson, S. 74, 75
Andrews, R. 94
Aptekar, L. 10, 49
Archer, M. 31
Atkinson, A. 18
Atkinson, P. 126

Barton, A. 28, 30
Bates, F. 33, 56
Baudrillard, J. 37, 102, 103
Beck, U. 9
Bee, P. 109, 110, 208
Benthall, J. 36–7
Benton, T. 36, 40
Berger, P. 35
Bernard, H. 125
Best, S. 34, 36
Birkland, T. 22, 78, 80, 81, 225
Blaikie, P.: and disasters generally 2–4, 9–10, 15, 21, 28, 39–43, 47, 56; and research on Northridge communities 105, 219–21, 223, 228; and vulnerability, sustainability and social change 230, 232, 233, 234

Blocker, T. 33
Blumenberg, E. 49
Blumer, H. 29, 30, 35
Bolin, R. 83, 118; and disasters generally 12, 32, 33, 38, 45–9 *passim*, 59–60; and responding to Northridge earthquake 166, 175; and restructuring after Northridge earthquake 185, 204; and vulnerability, sustainability and social change 219, 220, 226, 232
Bolt, B. 16, 17, 19, 65, 66, 70, 78
Bolton, P. 30, 44–5, 49, 59, 88, 144, 219
Booth, D. 42
Bosworth, S. 62
Bottles, S. 72–3, 232
Bourdieu, P. 31
Brookfield, H. 39, 42
Brown, E. 41
Brown, L. 2
Bullard, R. 9, 46, 51, 63, 219, 230, 231, 235
Burton, I. 30, 42, 46–7

Camacho de Schmidt, A. 57–8
Cannon, T. 5, 6, 9, 21, 28, 42, 45, 46, 51, 53, 219
Chambers, R. 10
Chomsky, N. 37
Cisneros, H. 94, 156
Clinton, W. J. 94
Cobb, J. 231
Cockburn, A. 231
Collins, F. 53, 195, 221, 229, 237

SUBJECT INDEX